X-RAY VISION

X-RAY VISION
A Way of Looking

RICHARD M. SWIDERSKI

Universal-Publishers
Boca Raton

X-Ray Vision: A Way of Looking

Universal-Publishers
Boca Raton, Florida • USA
2012

ISBN-10: 1-61233-108-4
ISBN-13: 978-1-61233-108-9

www.universal-publishers.com

Cover photo © Bendao | Dreamstime.com

Publisher's Cataloging-in-Publication Data

Swiderski, Richard M.
X-ray vision : a way of looking / Richard M. Swiderski.
p. cm.
ISBN: 978-1-61233-108-9
1. X-rays. 2. Science in popular culture. 3. Science fiction—
History and criticism.   I. Title.
PN3433.6 .S95 2012
306.4—dc23

                                            2012907654

ANTICIPATION OF X RAYS.—In Thomas Nugent's translation of Father Isla's 'History of Friar Gerund de Campazas,' published in 1772, we are informed that there is a popular idea in Spain that certain persons called Zahoris are "born with the faculty of seeing clearly anything which is covered, even though it should be under the earth, so that it be not covered with a blue cloth" (vol. i. p. 365). K. P. D. E.

† A popular idea, that there are perfons (whom they call *Zahoris*) born with a faculty of feeing clearly any thing which is covered, even though it fhould be under the earth, fo that it be not covered with a blue cloth.

TABLE OF CONTENTS

LIST OF ILLUSTRATIONS

INTRODUCTION

X-ray vision created a public vision and the threat of being seen. But X-ray vision never has existed. Why do we think it might? The spread of surveillance that accompanies X-ray vision intimates the all-seeing eye and recalls the anxious subjectivity of the observed viewer.

The discovery of X-rays did not introduce a new way of seeing. It introduced a new way of looking. News of the discovery spread more quickly and more widely than the images themselves. The first newspaper notice of X-rays was accompanied by an X-ray photograph, of Berthe Roentgen's right hand, still the most reproduced X-ray image ever. The hand was a simple demonstration of the force of the method. Other medically useful images might not be so readily recognized. Artistic X-rays of animals, again obvious by the correspondence between external and internal structure now visible, made an occasional appearance in printed media. And of course there was the novelty of X-ray calling cards.

Most of those who learned of X-rays learned from written descriptions or word of mouth. They were known to reveal interiors living, dead and never living. The closest epistemology was that of anatomy, and the most accessible anatomy was that of the skeleton. X-rays had to spread out the bones of the body in their natural order. X-rays were rapidly and variously equated with a way of looking that already incorporated considerable information about the body's structural insides.

The knowledge of the anatomist and the physician was scrolled into the X-rays. Genuine X-rays are shadowgraphs that have to be interpreted by those knowledgeable both in the technology of image formation and the internals of the body represented. For most people X-rays were a form of light that lit up the body like an opening page of an anatomy book.

This awareness of X-rays fostered X-ray vision. It was a way of assimilating the technology to eyesight and of registering the fears that such an eyesight stirred. A case study of X-ray eyes and X-ray vision begins with an automatic assumption of how X-rays acted when directed at a body, and continues in social and cultural developments of that theme apart from the actual applications of X-ray imaging technology.

X-rays were assimilated directly to eyesight, and they were pictured in the process of being a form of eyesight. Seeing with X-rays was referred to the subjective experience of those who could do so, or who gave evidence of being able to see by X-ray light. The act of seeing with X-rays was itself imaged, as a form of illumination described as a normal act of sight extending inward, or a through a surrogate technology giving an impression sufficiently like the idea of X-ray imagery to take its place.

X-rays were a beam aimed at the person or object being X-rayed, without thought to the screen or photographic plate that had to be on the other side of the subject. They were taken to be like a directed beam of light aimed into the darkness to light up what could not be seen by present illumination. X-ray vision was beam-like looking parallel to the gaze, the stare, the inquiring eye. It therefore aligned with the existing social forms of the gaze and was fraught with danger beyond the inherent harmfulness of ionizing radiation. The evil eye, the voyeur and the spy were aligned with X-ray vision from the start. Its look was seen.

X-ray vision was and is an image-idea not strictly bound to the natural force and the technology that name it. There always is a sense of the elements of the idea, of being able to look through the barriers of bodies and objects that preceded the discovery of X-rays. This is evident from the residue of ancient vision beliefs and of the folklore of vision present in X-ray vision narratives. It is also evident from the occasional attempt to reconcile X-ray vision with the physics of X-rays, or to invent a dream physics that would make X-ray vision possible. Just as common are the explanations of why it would not be possible as a form of human eyesight. X-ray vision is an image-idea tending toward an image, which frees it from X-rays and returns it to the antique light of its real origins.

That antique light, which opened darkness of the earth and the body, was much sought in the early twentieth century, as night was made day and secret cells were mercilessly flooded with brightness. X-ray vision achieved that sudden illumination. It recovered the experience of first light abolished by electric light fixtures replacing gas light on city streets and in houses. As artificial light displayed more detail and greater expanse than it was comfortable to see all the time, so X-ray vision could not be shut off once it was aimed at its object. The solid bone structure promoted by anatomy liquefied before the X-ray eyes. Acquiring X-ray vision was a new caption for an old tragedy. The supposition of its presence was a new caption for an old comedy.

This book is an excursion into the manifestation and development of X-ray vision and X-rays eyes from their first naming to their widespread acceptance as an image-idea. X-ray knowledge really was a new use for an existing technology and was spurred in its course by medical, industrial and security needs. X-ray vision also spread rapidly, following pathways and currents traced in these pages.

The first chapter recovers two of the shaping antecedents of X-ray vision. In the dance of death and then the phantasmagoria, the skeletons of the dead summoned up for final judgment became living skeletons going about daily business. This display taken up by the cinema mimics a world made transparent by the fall of X-rays. The individual eye of the zahori saw into the earth to water, precious metals and graves with valuables. Both visions were summoned up anew by X-ray vision.

The first fictional registers of X-ray vision came within weeks of Roentgen's January, 1896 announcement. A retired British India administrator imagined how a retired British India administrator might concoct eyedrops that made his eyes X-ray eyes, and recoiled from the result, which he did not consciously equate with the living skeletons made by the famines passing through the country under British rule.

X-rays might be able to penetrate or bypass the normal route of light through the eye and carry shapes and ideas directly into the optic nerve. Seeing X-rays or seeing with them was a possibility to be explored. The rays beamed into the eyes of the sighted and the blind from Roentgen onward did or did not register visibly. Perhaps X-rays caused the interior of the eye to fluoresce and thus make objects held in the beam into visible shadows. No one who attempted looking into X-rays during this brief period of experimentation claimed that the interiors of objects were visible. What would always be understood as X-ray vision was not accomplished with X-rays.

An edgy play of the spirit began the new century's beliefs about seeing with X-rays, all the way through to unpleasant truths. X-rays must be part of the light and vibrations always present and visible with the right apparatus. X-ray opera glasses were rumored to be available: the audience in the theatre and the crowd on the street would be stripped bare in the new light. Legislation was proposed, impenetrable underwear was advertised, perhaps as a joke, to stop this ocular invasion of privacy. Edison was supposedly about to issue X-ray spectacles. Cartoons made light of the look through the skull into the very thoughts of another. Even the blind might see if X-rays were beamed directly into their eyes.

A pacifist artillery officer imagined the next step in X-ray vision: the evolution of a being with wooden eyes, who therefore would see only by the ambient X-rays. The fanciful portrait of the Xylope's beloved, a romance become gelatinous, brought exclamations of disgust from commentators who stopped at that. A French socialist doctor cultivated a working class woman's ability to see through paper, which was taken to be X-ray vision after X-rays were discovered.

X-ray eyes were claimed by women who earlier would have been labeled medical clairvoyants. The eyes themselves with a little training saw as an X-ray apparatus did, and the claim was verified by the diagnoses the "girl with the X-rays eyes" uttered as she looked into her client. The look of the eyes, pupils dilated by belladonna, was sold in X-ray beauty parlors together with the ability of those eyes to see the innermost desires and troubles of future girls with

X-ray eyes. The inner experience of an X-ray eyes beauty was expressed only when a fictional plastic surgeon of body and soul invented Roentgenol and allowed a young woman to see the void in the heads of the rich, whom she joined.

The see-through abilities of the eyes alone merged briefly with the tradition of the gifted dowser-zahori appearing briefly in the United States-Mexico border states. The X-ray rubric conveyed news of one young man's ability to see water and valued fluids within the ground among hopeful entrepreneurs across the nation. A South African man possessing comparable talent rejected the X-ray eyes designation of an American journalist, but was proclaimed under that title just the same.

The inclination to emphasize the eyes over the X-rays continued with other fictions that began as the adventure of a man given cosmetic radium treatment and able to generate the rays from his brain. The man who could see through walls and detect plots on the other side became in the passage from France to America the man with the X-ray eyes illustrated with beams streaking out to render evil secrets plain to his view. The viewpoint of X-ray eyes was ambiguous: the unusual sight for him is usual for those he spies upon. Outsiders must see the process of viewing, the wall dissolved as well as what is happening on the other side, for the viewer truly to have X-ray eyes.

The individual gift of eyes able to see vibrations of force above and below visible light would uncover forms of life that move about unseen, much as scientists with their microscopes and telescopes detected extraordinary invisible everyday creatures and energy centers. These other worlds accessible to those with radiant vision remained potentially present in any view. Instruments that gave access to these nearby worlds were imagined as X-ray remote seeing and magnifying devices. They eventually were realized as television and nanovision, at first assembled from X-ray apparatus components.

The ability to see with X-rays alternated between a personal endowment and an instrumentality. At first the personal endowment replicated the capacities of X-ray equipment, but being human was superior. Young Leo Brett, the son of a physician was able to see into bodies from an early age when his father (and he alone) placed him in a hypnotic trance. Leo saw anatomical diagrams inside people, much clearer than X-ray photographs and in color. Beulah Miller from childhood saw into playing card hands and people's pockets, a claimed clairvoyance labeled X-ray vision.

Tests administered to Leo and then Beulah by a scientist of phenomena assumed to be psychic and then by an early experimental psychologist were the source of a slightly sustained newspaper renown for them as the psychic scientist tried to preserve his X-ray vision contention against the psychologist's materialist conclusions. While the psychologist's view has prevailed, that Beulah's vision was a matter of reading subtle cues, her performances of detailing concealed markings and objects became the standard for individual X-ray vision.

Houdini readily unmasked a Spanish X-ray vision performer in the 1920s, which did not discourage others from attempting to advance their acts. The Kashmiri virtuoso Kuda Bux, who shifted from firewalking to blindfolded readings, spanned the variety stage and early television with his acts of self-proclaimed X-ray vision. His technique never was revealed. He attributed his abilities to spiritual discipline, which he offered to teach without anyone taking him up on the offer. Other Indian yogis included X-ray vision among the powers their teachings could induce. Easier to copy were the packaged magic tricks that democratized X-ray vision in popular manuals.

X-ray eyes, like the title "Chief", were imposed on the major league baseball player of Chippewa descent, Albert Bender. The pitcher's familiarity with the preparations for the pitch enabled him to read the signs given by the opposing team's pitcher to the catcher. This wasn't called sign stealing or considered unacceptable at the time as it is now. A baseball writer combined Bender's insight with his dark eyes and his Indian background to give him a form of individual X-ray vision and X-ray eyes. His sight was visible to others and so were his eyes. Like other professional attributions this was applied to few other ballplayers after the unique instance.

The shower of ambient X-rays exposing the innards of buildings and at times people in the form of X-ray fashions, was sustained as a transparent world exposed at times in clothing design, painting, architecture and films. A few artists, Frantisek Kupka and Naum Gabo, for instance, fashioned their works as X-ray passages through matter without claiming X-ray vision. Houses and rooms losing their walls was a standard of cinema not ascribed to the activity of X-rays or the audience's ability to envision them. Architects designed glass houses generally without an X-ray reference. The filmmaker Sergei Eisenstein sketched an X-ray inspired film project named *Glass House*: a building entirely of glass where human cruelty was acted out with no barriers. The screenplay never was completed nor has there been any such film. The considerably less ambitious reality television series that unrelentingly follow people in their struggles for acceptance and cash don't look through walls.

X-ray vision and the transparent world came together in the figure of the superhero, an extraordinary individual whose powers had to be viewed in their exercise. They were viewed first in comic strips, which show Olga Mesmer, a product of her Venusian mother being exposed to X-rays by her scientist father, looking with a burst into walls but not what she sees on the other side.

Superman did less when he first made use of his X-ray vision in the early years of his comic. He only alluded to what he saw with his "X-ray eyesight".As the strip continued readers see inside enclosures as Superman does often with beams from his eyes breaking open the barrier. His vision acquires the X-ray accompaniments of telescopy and microscopy with the addition of heat vision.

The spectrum of media hosting Superman, his companions and antagonists, were a site for old and new forms of the transparent world as techniques improved. The method of simply running footage of what Superman saw with his X-ray vision was replaced by a return to the phantasmagoria and the anatomical diagrams of earlier transparency, following by moving X-ray like shadows. Successive iterations of the Superman type gather the modes of envisioning X-ray vision into diverse packages.

X-ray spectacles were revived in the 1950s using as a surrogate diffraction imaging which had been patented decades earlier but never successfully commercialized. These cardboard spectacles sold cheaply in comic books and popular science magazines were both to see and be seen (one design had swirling spirals around the peephole). The immemorial X-rayed hand appeared in some ads, a woman silhouetted inside her dress in others. The X-ray specs were an invitation to baiting girls with the signal that juvenile voyeurism was going on. They contributed a field of nostalgia for those who recalled they actually worked.

Like the later Superman spectacles the film *The Man with the X-Ray Eyes* reviewed all the forms of public X-ray vision, but in the tragic mode, from the individual point of view of a doctor who experimented with eyedrops. This film added the X-ray spectacles diffraction grating vision to the implicitly naked bodies, the skeletons and anatomy diagrams. Individual X-ray vision reviewed and replaced the transparent world of phantasmagoria, X-ray skirts and glass houses seen by all.

Men who witnessed nuclear and thermonuclear test blasts unintentionally attained the truest X-ray vision when they raised their hands to shield their eyes, yet they did not use that phrase to name what they saw. X-ray vision was extended from the X-ray realms to peoples who did not themselves evoke the label. Western commentators saw both individual X-ray vision and the transparent world in social practices and imagery of the Hopi, !Kung, Maya, Yoruba, Akan, Wana and Ojibway among others.

X-ray vision and especially the Superman model is used to introduce technology for high degrees of magnification such as dichroic microscopes and the Chandra X-ray observatory. The very small and the very remote once believed to be accessible through X-rays finally were accessed through X-rays and acknowledged as such, though the object seen does not resemble the object imagined. X-ray vision as attained is not X-ray vision at all.

The flux of X-ray vision as imagined is replayed in the stories of tabloid newspapers that recast it in fictions reflecting current news stories. The force of lightning imprints the capacities of an infrared surveillance camera on the brain of a police officer. He solves crimes, locates lost keys and celebrates the long-standing voyeuristic potential of his gift. The other tabloid accounts have a similar bent whether through eyeglasses or special abilities except for the one reporting at third hand about the children living in a village near Chernobyl after the meltdown of nuclear reactor.

The voyeur content of X-ray vision dominates the next surrogate technology, infrared video cameras that are able to record the skin within the clothes. They are removed from the market for publicized reasons, increasing the sales of similar cameras and the appeal of websites and booklets that instruct users how to modify the present camera for daytime infrared recording. This is an attempt to privatize the transparent world and mechanize the X-ray eye to make it universally available. But then, it only works for you if you think it does, like all X-ray vision.

X-ray vision and X-ray eyes passed through phases of simulation and acceptance. A parallel history of X-ray simulation technology accompanied improvements in X-ray imaging, free from constraints on the authentic technology necessitated by its demonstrated deadliness.

Diffraction grating, bird feathers, infrared filters and ultrasound all provided substitute visions safer than genuine X-rays. X-ray vision technology underwent its own evolution independent of X-rays and their imaging.

The interiors X-ray eyes saw were not X-rayed interiors. The problem was put aside by those who performed having X-ray eyes. They simply reported information that could not have been delivered to them by normal senses. As X-ray vision ceased to be a novelty it became a party game reduced to a set of rules for general consumption. That only encouraged the emergence of a virtuoso or two.

The mystique of unresolved social insight could not be dismissed so easily. The ability to look into people and enclosures, whether explained by a pseudo-technology or not, was impelled by the anxiety of that vision and framed by the regret of having used it. In an age of secrets and thickening walls around them X-ray vision was compensation for the inability of most people to breach those walls in any other way. For the inability of humans to change bodily in tandem with their technology.

With time X-ray vision aggregated many themes of extraordinary powers focused on entertainment. New devices for seeing the very small and the very far made use of the metaphor, X-rays themselves or both. Ideas of how we see had to make their apologies to X-ray vision. New devices permitting X-ray vision were invented in that space where their success cannot be denied because it has been purchased.

Having adapted old beliefs in penetrating vision to the new technology, X-ray visionaries assumed that a technology could be created to fulfill the beliefs.

X-ray eyes embody a longing for universal surveillance, a force in itself, and a fear that universal surveillance might be achieved. They are an index of the growing subscription to surveillance that should be universal. X-ray eyes are sent away always looking back, like an abandoned pet on the road behind us or a probe headed toward other galaxies.

1

ANIMATED SKELETONS AND THE
TREASURES OF THE EARTH

The world-wide spread of X-ray knowledge in the early twentieth century was the culmination of a development that began with the magic lantern. X-ray photographs displayed the interiors of animals, humans and objects while light-based photographs (and of course sight itself) displayed their exteriors. Light projected solid and moving forms; X-rays projected static shadows. X-rays were a disadvantaged marvel.

There was Thomas Edison's fluoroscope first exhibited at an exposition of electrical technology in 1897. The passing audience watched a screen as they extended their hands to the other side and saw the bones inside the flesh. Individual entrepreneurs sold the opportunity to look into a hand or shoe. Hands are flat and easy to read. Disclosing their bones was the inescapable subject of early X-rays. An X-ray projection spectacle of the entire body was easy to imagine. It left the living body open. The skeletons of the past gained new life.

Moving skeletons were sometimes realized in magic lantern shows. Christiaan Huygens, the main inventor of the magic lantern, was intrigued by Hans Holbein's animated skeletons in his series of engravings *The Dance of Death* (1538). The skeletons dance and caper as they interrupt bishop and pauper in their activities and lead them off to death.[1] Huygens painted the skeletons life-size on a wooden fence in his yard. He also made a series of ink sketches of standing skeletons (1659), their multiple limb positions and dotted lines indicating motion.[2] One of the skeletons bends forward as the

[1] Holbein (1971)
[2] Mannoni (2000: 38-39)

21

lines seem to indicate his skull has fallen into his right hand; another tosses the skull upward as he bends back, his left hand rising to keep his own skull from falling.

Laurent Mannoni theorized that Huygens meant the drawings to be on separate glass sheets, one slid over the other while the projector's light beam shines through them. Sequence of slides or articulated parts conferring motion became a practice of projection mechanics. Huygens himself used his projections to amuse friends and family. In the hands of professional projectionists who perfected the equipment and slides during the century that followed the magic lantern made its way around the world.

The early cinema reflected Huygens' wish to make the skeletons move. A 45 second short by the Lumière brothers, *Le Squelette Joyeux*, *The Happy Skeleton* puts the bouncing puppet through many of the same paces Huygens sketched out almost 250 years earlier. The skeleton drops and recovers an arm and a leg. He collapses in a heap but his legs rise dancing. Erect again, his head flies off and returns, and he finishes his dance with a few high music-hall kicks. The dance of life preserves the dance of death.

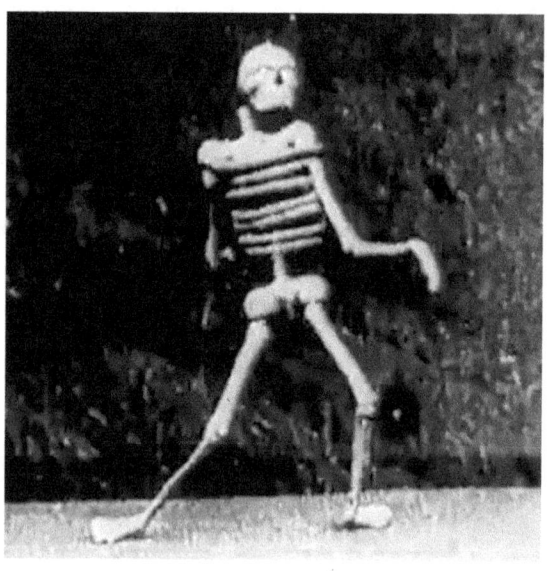

1. Still from Le Squelette Joyeux (1897), Lumière

Lantern shows were genre projections of everyday life, of quarrels and romance, and educational projections of plants and animals, butterflies, worms and tadpoles, some of them living as long as they could survive encased between glass plates with a light beam passing through them. Liveliness again betokened death. Apparitions and skeletons emerging from crypts remained part of the repertory.

Etienne-Gaspard Robert, best known as "Robertson," an innovative organizer of projection events during the late eighteenth-early nineteenth century, claimed expertise in "phantasmagoria," "a science which deals with all the physical methods which have been misused in all ages and by all peoples to create belief in the resurrection and apparition of the dead."[3] Robertson was the forerunner of other magician exposers of frauds upon the credulous, Robert-Houdin and Houdini, to name two-who made the exposure part of their own deft shows achieved entirely by human means.

By abjuring miraculous powers Robertson freed himself from scrutiny by church authorities and laid a claim, not generally accepted, of being a scientist. His technical innovations included projectors that could be moved to enlarge or reduce an image while keeping it in focus, well suited to form environments of the dead rising and spirits gathering in capacious darkened spaces, as the chilling glass harmonica sounded and powerful artificial winds blew.

Religious background and gothic literary atmosphere predisposed his audiences to react spontaneously to the rising skeletons and flying demons while conscious it was a paid performance. The recent overthrow of the French monarchy had worked to Robertson's advantage, increasing anxiety and making religious properties available for secular entertainments. An often reproduced engraving of a 1797 phantasmagoria portrays the audience seated in pews raising their arms at the approach of the winged death's head and hovering hermaphrodite. One man reaches for his sword, another cowers on the floor, as the censers pour out smoke.[4]

Robertson kept his apparatus out of sight. It was the mechanism that had become esoteric.

---

[3] Quoted by Mannoni (2000: 148) from Delrée (1954: 19)

[4] Robertson (1834)

2. Phantasmagoria of Robertson in the Capucine Cloister, 1797. Robertson(1831: frontispiece)

Among Robertson's surviving painted glass plates are a death's head image between two wings that could be worked to flap, a shrouded skeleton that is housed in a crypt in one slide and stands on a nearby tomb in another, a skeleton holding the hourglass and scythe of Death, and a skeletal rider seated atop a skeletal horse, one of the Horsemen of the Apocalypse. In the static glass paintings they still appear to be arriving from the beyond. [5] An 1840s version of Robertson's Fantascope projector owned by a Belgian collector includes a projectable mechanical puppet of a skeleton that turns its head and moves its mouth as the operator cranks a handle.[6]

After Robertson the funereal and otherworldly subjects of projection shows diminished in number as the techniques of imparting motion to the figures proliferated. Whenever active skeletons appeared they were the reanimated dead, not skeletons of the living. The phantasmagoria mode of displaced necromancy persisted in the successor of the magic lantern shows, the cinema, from the many skeletons of the early trick filmmakers to the musical skeletons of Disney's 1929 Silly Symphony, *Skeleton Dance*, to the warrior skele-

---

[5] Reproduced in Remise, Remise and van de Walle (1979: 47, 50 and 53)
[6] Described by Mannoni (2000: 156-57)

tons of Ray Harryhausen's figure animation for *Jason and the Argonauts* (1963), to the still-corroding stop-motion characters of *The Corpse Bride* (2005).

X-rays did not expose full skeletons, but death and the cinema did. An eye able to see with X-rays as with light might also see live, moving skeletons. Independent of the skeleton show and converging with it under the X-ray beam was a penetrating eye and a way of looking. The alignment of this eye with the skeleton show was a promise X-rays threatened to keep. It was a promise made long before X-rays were known, long before the phantasmagoria was staged.

Pedro de Hoyo, from 1556 to 1568 head of the authority servicing the royal residences in Philip II's Spain, wrote to the king expressing enthusiasm about a young boy who was able to see water under the earth.[7] Only able to work on days that are sunny and bright, this boy was the best *zahorí* de Hoyo had yet employed, and he certainly would find the needed water resources. The king's scrawled reply preserved in the Spanish state archives gives royal assent to the project, but de Hoyo's account of the results is fraught with disappointment. These *zahorí*s are not always right, he dejectedly recorded, after the boy kept urging the official to have his men dig deeper at a spot where no trace of water appeared.

The king may have lent his authority to this *zahorí*'s search, but others making the same claims of seeing water, minerals, and the buried dead within the earth were interrogated and sentenced by the tribunal of the Inquisition to corporeal punishment and improving instruction. The Jesuit theologian Tomás Sanchez allowed that *zahorí*s might be harmless fakes who if they did succeed in locating earthly riches were under demonic influence, their immortal souls imperiled. Benito Feijoo, a Benedictine theologian, dismissed the zahoris' claims because light cannot penetrate the ground.[8]

A celebrated zahori named Pedegache in Portugal was revered for her eyes' sight and the prosperity she brought to treasure finders. A Portuguese woman in seventeenth-century colonial Peru

---

[7] Goodman (2002: 17-18)

[8] Feijoo (1739: 2, 325-26)

"could see right into people's insides" and diagnose their ills, as well as find underground resources.[9] This María Martinez admitted to an inquiring priest that her imported powers had to be amplified by more powerful artifacts from Peruvian Indians.

Fixing his own enlightened eye on the zahoris, Gilbert Charles Legendre discovered that they were peculiar to the Iberian Peninsula, Spain and Portugal, the belief in such people having been brought there by the conquering Moors.[10] Pierre Bayle's *Dictionnaire General* substantiated this identification and the 16th century French and Spanish sources that Legendre cited. The association with Arabic was confirmed by the derivation of the word from *zahur*, "a lie, zuhou means a cheat, from which zahouri has been made."[11] That is not the lexical meaning of the word's root, and no reference to *zahur* and its derivatives in Arabic dictionaries refers to water-sighting or cheats.

The word at base means "to shine, glow or bloom" with numerous corresponding derivatives as the root is modified phonetically.[12] *Zahori* was a Spanish word derived from Arabic for people who brought to light what is hidden within the earth, whether corpses, treasure or water, or who deceptively claimed to be able to do so. Where claims were made and stories told about zahoris, they were successful; where they were put to the test by those hoping to take advantage of their skills, they invariably failed. The term began as a promise and ended as an accusation, which was its state when English writers recorded it during the early 19th century.

W.F. Barrett, late 19th century psychical investigator, appended a discussion of zahoris to his part-empirical investigation of "the so-called divining rod."[13] He cited the 16th and 17th century sources of Legendre and Bayle, and only added a contemporary Swiss observation that every village in Switzerland had its water-finder who could see within the earth.

---

[9] Silverblatt (2000: 118-19)

[10] Legendre (1735: 500-1)

[11] Weston (1810: 780)

[12] E.g, Wehr (1975: 384)

[13] Barrett (1897: 367-72)

While there were no contemporary zahoris in name or deed, the Arabic root of their name, "bringing things to light", connected them to "the ability to detect hidden objects in some transcendental manner" attested for certain persons in many countries. Barrett saw the same root in the title of the Hebrew mystical treatise, *The Zohar*. The zahoris were therefore assimilated to other seers and finders of concealed treasures, and separated from the reputation for fraud that had attached to the name in recent usage.

The zahoris and their ancient Roman counterparts, the *aquilegos*, were among the many types of local water and treasure finders who lingered in scholarly awareness and popular consciousness at the beginning of the twentieth century. In places where water was sorely needed, that is, dry lands where European farming and cattle-raising enterprises were being extended, the water finders did reappear. Zahoris, in deed if not in name, reappeared when petroleum became an equally desirable fluid to be seen underground.

The discovery and promotion of the X-ray vision of matter intersected both with the lingering phantasmagoria, where X-rays seemed to be showered on people, displaying their moving skeletons, and with the zahori traditions of sight into the ground. Where they were evoked, X-rays conferred a novelty on these ground and body peekers and quit them of the cast of deception they had previously earned. The spread of spiritualism, which reconstructed folk beliefs within a scientized supernaturalism, contributed to the passing invention of an X-ray spiritualism.

One new element was the inescapability of what X-rays revealed to the eye, and the phantasmagoria that might suddenly come into view where treasure was expected.

2

ROENTGEN'S CURSE

Within two months of Roentgen's January, 1896 announcement of X-rays an English literary magazine published a story featuring a man who gave himself the ability to see the world through the new radiation. It seems natural to have imagined such an ability, but what do the circumstances of this first entry say about the ability itself?

The protagonist of the story, Herbert Newton, is an India civil servant who spent years "in the chase of the cholera microbe," and is now retired with wife and children to a country house in England. His wife's painting and photography pastimes engage him, and in the wake of the X-ray discovery he vows to pursue "the photography of the invisible." He bases his quest on realizing a homology.[14]

> If the invisible ultra-violet rays can be made perceptible to the eye by means of the fluorescence they excite in certain substances, then why should not other rays, such as Röntgen's X-rays, be made visible to the eye by some chemical means?

Newton prepares a liquid sensitive to X-rays and experimentally applies it to the inside of the eyelids of his dog Dash, who has remained faithful to him as his obsessive experimenting alienates his wife and friends.

The dog's cries of terror and flight into hiding upon opening his eyes do not deter Newton from brushing the liquid on the insides of his own eyelids. He is exhilarated by the prospect of seeing as with

---

[14] Crosthwaite (1896: 471)

the eye of the Creator, having the world transparent before him, being able to identify the causes of illnesses. He is not insensible to the likelihood of locating minerals and precious stones within the earth, and hidden ancient cities with their treasures. The zahori motif peeks in, followed by the phantasmagoria.

The sight of his wife, who has become a set of bones moving in an envelope of flesh, appalls and disgusts him. His reactions cause her to summon the doctor, who appears to Newton a skeletal caricature of himself, his pompous professional airs reduced to a collection of sliding bones. Newton cannot abide being surrounded by living skeletons who once were fully fleshed people. He takes to his bed eyes closed until the effects of the liquid wear off.

During his convalescence his wife clears out his laboratory and makes it into a billiards room where he might host more companionable after-dinner gatherings with his gentlemen friends. Newton rescues his notebooks and sends them to a German savant named Dr. Gleichen, who informs him that he has been able to replicate the formula and will test the liquid on his own eyes. At first Dr. Gleichen is as stimulated as Newton was by the prospect of seeing with X-rays, but his communications grow less enthusiastic, then cease entirely. Later Newton learns of Dr. Gleichen's death.

Newton's frenzied drive toward scientific discovery for the benefit of humanity and himself turns sour when he makes himself his own experiment. The ontology of the invention overtakes its epistemology and Newton succumbs to theatrical horror settling into a satirical melancholy. He doesn't give in to animal urges summoned up from within himself like Dr. Henry Jekyll (1886), nor does he go mad from the change to his worldview and envisioned power like Griffin, the protagonist of H.G. Wells' *The Invisible Man* (1897). He settles back into the attitudes of a late Victorian squire and plays a trick, possibly vengeful, on a German scientist whose name suggests an authentic contemporary, the prominent Berlin optician Dr. Alexander Gleichen. It might be an author's pun to use the name: "gleichen" means "to resemble."

Newton's dog, and then Newton himself, sees by receiving X-rays from a universal source projected through his surroundings onto the platinocyanate screen of his retina. He sees as by an extraordinarily sharp fluoroscope. The visibility of metal fittings is enough to allow him to navigate the transparent wood and stone of his immediate surroundings, but people's apparent loss of clothing and flesh thrusts him away from society. As the editor of *Nature* magazine remarked in an article that appeared the following September, the author of the story demonstrated a better understanding of the science than several other contemporary authors of science-based speculative fiction.[15] A truly cautious scientist, the editor added, would have painted the interior of only one eyelid rather than both.

The story is the first fictional projection of X-rays into a medium of personal subjective vision. This is what the world would look like if X-rays took the place of light. The ambition is cosmic but even at the domestic starting point the presumed blessing is a curse. The fright induced by the phantasmagoria carries over to the skeleton show of life illuminated by X-rays.

The author of this story was Sir Charles Haukes Todd Crosthwaite, KCBI, sixty-one years of age and in the first year of his retirement from the British India service, where he ended a career of steady ascent through the ranks of colonial administrators as Lieutenant-Governor of the North-West Provinces and Oudh. Crosthwaite was not involved in the search for epidemic disease pathogens like the protagonist of his story. Patrick Manson, who was, corresponded with Crosthwaite seeking his support for the work of Ronald Ross, who identified the malaria vector.[16] The microscopy that disclosed the pathogens of cholera, malaria and other diseases that troubled the colonies was the means of making the invisible visible for Crosthwaite's generation as X-rays were for the generation to follow. Microscopy and X-ray photography were projections onto the same screen.

"Röntgen's Curse" was Crosthwaite's first published work since his retirement. After he had been in the India service for twelve years he published a brief book, *Notes on the North-West Provinces of*

---

[15] Lockyer, ed. (1896: 454)
[16] Haines (2001: 112-13)

India (1869) discretely authored "by a District Officer".[17] The title was the same as an earlier book except that the previous author had named himself. Junior colonial administrators tried to attract attention by issuing informative publications on their territory and Crosthwaite's discoverable anonymity was gesture of authoritative modesty, not of self-concealment. *Notes* was a dry account of native soils, agricultural techniques and land tenure practices, clearly derived from observations but not from anything resembling participant observation. They are an inspection of the land with the eye for improvement in production sought by an official of the empire.

Crosthwaite's publications during the rest of his career in the British colonies of India, and Burma, where he was Chief Commissioner before assuming the lieutenant-governorship in the northwest provinces, were legal documents collected in administrative compilations. He seems to have been biding his time."Röntgen's Curse" clearly was a rapid response to current events. "Thakur Pertab Singh: The Tale of an Indian Famine", which appeared in an issue of *Blackwood's Edinburgh Magazine* the following year,[18] was a product of much longer consideration, the first element in a publication program that continued until the year of his death, 1915.

Crosthwaite's second published "tale" does not have the appearance of fiction. In the manner of his *Notes*, the text describes the land, only this time the land is afflicted by drought and the crops are failing. Thakur Pertab Singh, the descendant of a family of Rajput landowners who have seen their holdings dwindle under British rule, tries to maintain his dignity under pressure from a moneylender who holds a mortgage on some of the remaining land, and from the peasant tenants unable to meet their obligations. Most residents of the village of Gardanpur, and most of Pertab Singh's family, are forced to leave in search of food rations. Those who remain face a miserable death-a mummified corpse is dismembered by dogs-but Pertab Singh and his sons are able to weather the famine and even help the grasping banker fight off thieves who have invaded his home. With the new season the rains come and the

---

[17] Crosthwaite (1869)
[18] Crosthwaite (1897)

population returns to resume farming. Government remissions of taxes and other incentives help the landowners and people recover.

Crosthwaite's tale reflects the experience of one settlement during the "Great Famine of 18—" and the role of official policy in precipitating, and possibly in relieving the famine. The issue of government famine relief arose during the Great Famine of 1876-78, which at first was muted by the release of government stores of grain but then grew calamitous as relief was withdrawn. The total death toll was between 6 and 10 million people, the deadliest famine of the series during the British Raj. The famine led to the creation of a government Famine Commission. Nonetheless, there were several famines in different parts of India up to and after the publication of "Thakur Pertab Singh."

Crosthwaite was invested with the Order of the Star of India in 1888 as Knight Commander (KCSI), an honor that made him a life peer. The Order had been created by Queen Victoria in 1860 to honor (and secure the loyalty of) Indian rulers and British civil servants in India. Crosthwaite became a member of the Order in 1887, the year he was also named Chief Commissioner of the Burma colony, and he was knighted a year later, after he had played a key role in "pacifying" Burmese rebels. He witnessed a famine in Burma as well, during the years he was commissioner there, before becoming lieutenant-governor of the northwestern provinces in India.

As a member of the governing council of India after his retirement in 1895, Crosthwaite was engaged in policymaking. "Thakur Singh" refers to a famine decades earlier and the economic relations that made it and later famines come about. That story seems to be a soft nudge toward better anti-famine policy that Crosthwaite could not have advocated while still an India service officer. It was the first of several tales he published on Indian social problems. "Ai Kali!" on a plague in India, joined several other stories in a collection titled after "Thakur Singh" and published in 1913, two years before his death, the same year he published his account of *The Pacification of Burma*.

Other India service officers displaced their anxieties into fiction: Rudyard Kipling comes immediately to mind. The context of imperial domination and unprevented famine surrounded Crosthwaite's first story, seemingly irrelevant to anything personal beyond the conditions of his retirement. Crosthwaite had seen the human face, and body, of famine first-hand while he was in a position of responsibility. Thuttakadu Ramakrisha in 1896 published a book of poetry that included "Seetha and Rama: A Tale of the Indian Famine" containing the lines:[19]

> With hollow cheeks, sunk eyes and haggard faces,
> Like walking skeletons pasted o'er with skin

Ramakrishnan's verse tale was from the Indian famine, the same one that prompted Crosthwaite's tale. The famine earlier worked its way into his story of seeing by invisible rays, in the visions of "living skeletons" that drive Herbert Newton to his bed until he can open his eyes without seeing them. The prospect of relentlessly seeing with X-rays was horrifying for what it made impossible to ignore, or look at directly.

3. In a Famine Camp-A Pathetic Figure. Grant (1907: 93)

---

[19] Ramakrishnan (1896: 14)

A reflex that Crosthwaite developed during his years of India service, that facilitated his advancement, found its expression in a new force he learned about soon after he returned to England. If X-rays allow us to look inside, they also force us to. They are available to be adapted to an obligatory gaze we won't admit. No enlightenment follows Newton's attainment of this vision, only billiards.

3

SEEING (WITH) X-RAYS

Scientists, journalists and other commentators scolded the public for assuming X-ray vision was possible. There would have to be a source of hard X-rays, and the retina of the eye would have to catch them after they passed through objects. The world seen with X-rays was only a world imagined seen with X-rays. Yet from Roentgen onward there were those who said that X-rays made an impression of light in the eye. Along with the phantasmagoria and the zahoris this formed part of the background of the cultural phenomenon of X-ray vision and X-ray eyes.

"The retina of the eye is not sensitive to the rays," Roentgen wrote in his first published paper.[20] The eye observes nothing though its tissues are transparent to the rays. Roentgen had just reviewed at the beginning of his paper the substances that form dark on light shadows on the screen when passed between the ray source and the screen. The retina of the eye did not receive the same light impressions as the screen.

A few months later the physicist G. Brandes recounted some experiments in the minutes of the Berlin Royal Academy of Sciences.[21] Facing an X-ray source in a darkened room Brandes experienced a blue-gray glow that seemed to come from within the eye itself exclusive of the lens. Glass lenses placed before the eyes blocked the X-rays and extinguished the glow. A young woman without a lens in one of her eyes sensed X-rays through that eye but not as well through the eye with the lens intact. Brandes had set up

---

[20] "On a New Kind of Rays" IN Barker (1898: 5)
[21] Brandes (1896)

one inquiry in direct X-ray sensing: does the crystalline lens block X-rays?

Roentgen took Brandes' findings in stride and reevaluated observations he made during his first period of experimentation.[22] A feeble sensation of light he had over his entire field of vision when he faced a wooden door on the other side of which was aimed an active Hittorf tube must have been light generated by the X-rays within his eyes and not a deviation due to error as the cautious experimenter previously thought.

Figures of light became visible when he moved a piece of metal with a narrow slit cut through before his eyes. It even seemed that the glow of the platinocyanate screen that initiated the discovery of the new rays was the result of the rays themselves activating the sensitive parafoveal region at the corner of his eye and not the faint glow of the screen head on.[23] Roentgen himself saw everything anyone was ever going to see of X-rays directed into the eye.

Those who stood in darkened rooms in the presence of charged vacuum tubes saw X-rays even as they saw on the screen the shadows of objects and human limbs that X-rays passed through. The glow that suffused even surrogate X-rays originated in the glow of actual X-rays in the eye.

Thomas Edison's demonstration of a fluoroscope at a New York exhibition a few months after the initial announcement placed X-ray entertainment alongside the serious medical uses that also had been initiated. Members of the passing public could see the bones of hands they held on the irradiated side of a standing screen. Edison offered to loan a fluoroscope to physicians who wanted to use the rays to treat the blind, but made no claim of its effectiveness.[24]

"Blind Made to See" headed brief items in a number of newspapers nationwide in November, 1896. Lucien Bacigalupi, a San Francisco boy blind since youth, was reportedly able not only to see but

[22] "Further Observations on the Properties of the X-rays," IN Barker (1898: 7-8)

[23] Frame, Paul. "Wilhelm Roentgen and Invisible Light." http://www.orau.org./ptp/articles.orig/invisiblelight.ht.html

[24] McPartland (2006: 400)

to see into cases and enclosures with the aid of a fluoroscope "given" to him by Dr. Waverly Clarke. Lucien's father Peter was the proprietor of an Edison phonograph shop and kinetoscope parlor which included an Edison "X-ray viewer."[25] Dr. Clarke theorized that Lucien was blind due to a film over his eyes which the force of the X-ray light penetrated.[26] He claimed to be the first to induce sight in the blind with X-rays, bringing a testy refutation from Edison, who had seen one hand through his other hand placed over his eyes.[27] Edison explained that the blind would not be able to see with a fluoroscope since it projects only visible light. Lucien Bacigalupi spent the rest of his short life trying to earn a living as a concert singer and force his upwardly mobile father to support him.[28]

Newspaper reports of the blind seeing X-rays and regaining sight after being treated with X-rays spurred further research. An English chemist, G.H. Robertson, who had become blind as the result of illness, allowed a Crookes tube to be aimed into his eyes.[29] He felt a pulsation in his eyeballs which increased or decreased in intensity with the placement of the tube. A metal sheet put between his face and the tube ended the pulsation. He felt discomfort after a time, and had more of a sensation of light from the spark of the apparatus than from the tube. The technician told him that other people felt the pulsation in the eyeball but none had seen light. Robertson was in a sense relieved.

> As far as I am concerned, if I am to remain in a
> world of gloom, I prefer that it be untenanted by
> shadowy skeletons, and this is all that one could
> perceive by aid of the rays, even if the mere tech-

---

[25] Visible in a contemporary photograph, Phillips (1997: Illustration 39)

[26] Science and the Blind, *The Chautauquan*, August, 1896: 401. Includes excerpts from newspaper reports on Edison and Clarke nationwide.

[27] Glasser (1993: 306-7)

[28] Bacigalupi's Blind Son Dies in Poverty, *The San Francisco Call,* February 16, 1907: 4

[29] G.H. Robertson (1896)

nical difficulty of constructing an apparatus can be
overcome...

Robertson allowed that some blind people might be willing to accept that world in order to see, but he was not among them.

James Richard Cocke, who had been blind from infancy but learned the medical trade and gained a reputation as a therapeutic hypnotist, also felt a quivering or vibrating motion when the X-ray tube was focused at the back of his head, then his forehead.[30] He wore calcium tungstate lens glasses. When a board with brass letters was passed between the tube and his eyes he noticed a difference but could not read the letters because his only knowledge of them was tactile. A cylindrical object gave him feelings he never had before, what he described as "internal tactile sensations." He did not perceive this and other objects as realities external to the brain.

The author of the brief notice on the experiments with Cocke observed that blind subjects with a degenerated or destroyed optic nerve did not have thought-sight impressions like those Cocke described. If the nerves were intact as they were for Cocke, then the impressions came through. Unless there were new discoveries, he doubted that there would be much progress toward the blind truly seeing by way of X-rays.

The ability of the X-rays to transport matter, sensations, ideas, into the body and especially into the brain had been suggested by experimenters during the frenzied months after the rays' discovery was announced. The experiments with Henderson, Cocke and others followed experiments by Nikola Tesla exposing his own head to X-rays for 20-40 minute periods, with the primary intention of making an X-ray photograph of its interior. Tesla noted "a tendency to sleep and the time seems to pass away quickly."[31] If these effects are verified by keener observers, he exclaimed, he will believe in the existence of "material streams penetrating the brain." This might be a practical means of projecting a chemical into any part of

---

[30] Peck (1897)
[31] Tesla (1896)

the body, eliminating the need to use a hypodermic syringe for the delivery of medicines.

This view was compatible with Tesla's idea of the effects of X-rays, which he saw as carrying off particles of the target screen with great speed and force. The sensations were verified by the blind experimental subjects, but the materialist theory of X-ray action never was supported. Charles-Edouard Guillaume commented dismissively on Tesla's theory in a footnote to his physics text on X-rays and photography that same year.[32] Yet such was the enchantment of this notion that Guillaume soon published a brochure contending that X-rays could enter and be imaged by the optic nerves of the blind, perhaps leading to a form of eyesight.

Guillaume returned to his own interest, establishing reliable measurement standards, which would lead to the award of the Nobel Prize in 1920 for his discovery of a nickel-steel alloy with a very low coefficient of expansion. His X-ray enthusiasm was taken up by others, apparently unaware of the London and Boston blind sight experiments with X-rays. A venturesome newspaper, *The San Francisco Call*, organized a test of Guillaume's speculation with the assistance of Alexander van den Naillen, the founder of a Bay Area engineering school.[33]

They obtained from a slaughterhouse fresh calves' eyeballs, some of them "blind" and others "normal". Using a Crookes tube and generator loaned by van den Naillen they exposed both types of eyeball to X-rays. The blind eyeballs transmitted the X-rays, leaving the photographic plate on which they rested exposed; the normal eyeballs obstructed the X-rays, leaving an unexposed area on the photographic plate in the shape of the eyeball. The paired prints accompanying the newspaper article traced the travel of the X-rays down the optic nerve attached to the blind eyeball (left).

The newspaper concluded that Guillaume was correct and those blind because of damaged eyeballs would be able to see X-rays, as long as the optic nerve was intact. The eyeballs of normally sighted

---

[32] Guillaume (1896:102n)

[33] The Latest Phases of X-Ray Investigation, *The San Francisco Call,* April 25, 1897: 19.

people contained components such as the lens that obstructed X-rays while permitting the passage of visible light. X-rays entering the optic nerve directly would be visualized by the brain.

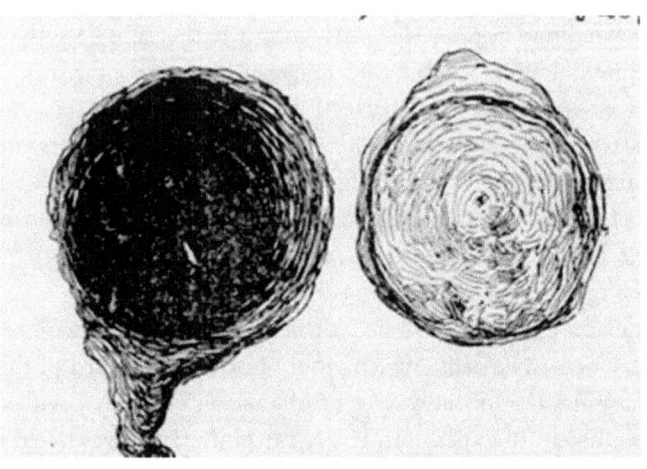

4. "Blind" calf's eyeball (left) and sighted calf's eyeball (right) exposed to X-rays while on photographic plate. Fn33.

No blind subjects were tested. It was difficult in practice at the time to distinguish between damaged eyeball blindness and damaged optic nerve blindness, and to find people who theoretically would gain sight from X-rays. The journalist-experimenters never reached the stage of learning the reactions of blind scientists like Henderson and Cocke to being irradiated with X-rays. Nor did they refer to similar tests in Europe that reached similar conclusions.[34] They did reinforce what was becoming an enduring notion in the face of scientific and personal testimony to the contrary, that X-rays could enter the brain directly and carry with them images, ideas, even matter. Not seeing X-rays did not preclude X-ray vision.

---

[34] La sensibilité de l'oeil aux rayons X, *Archives de l'électricité médicale* 5(1897): 520-21

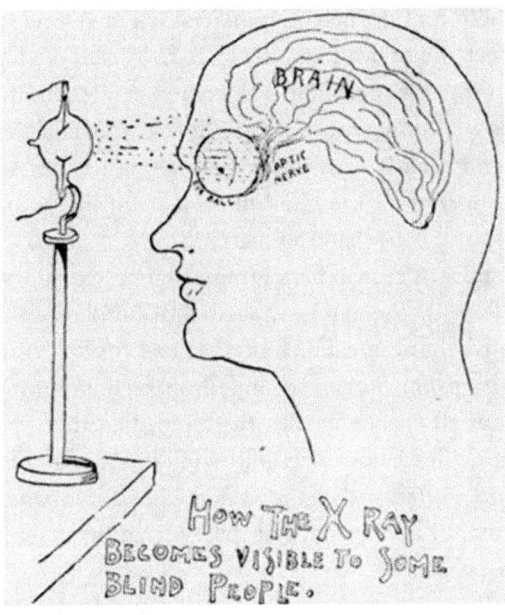

5. How the X-ray becomes visible to some blind people Fn33.

With the thought that if a variety of blind youth were placed one by one in a darkened room with a cloaked Crookes tube operating, those who could see invisible rays (cathodic, X, fluorescent) would be revealed, Dr. Foveau de Courmelles selected 204 from a school for the blind.[35] Some saw all or two of the three radiations. One who saw X-rays and cathodic rays said they were "red like the sun." The nine who saw only X-rays had damaged peripheral vision and could detect a little light. How the rays appeared to them is not described. Others with better sight saw nothing. Foveau concluded

> Impossible to reach definite conclusions in this work in spite of the large number of blind people observed. It appears to be, however, that the retina can acquire, in certain cases of blindness, a hyper-acuity comparable to the sensitivity of the photo-graphic plate that X-rays can expose.

---

[35] Foveau de Courmelles (1898)

Testing different forms of blindness yielded one combination of factors that might lead to seeing X-rays. The retina can *acquire* a sensitivity to X-rays under a specific set of conditions. This narrowed down Guillaume's criteria from an eyeball impenetrable to light to one slightly sensitive. The matter was not settled, but further mass testing of the blind did not seem likely to yield concrete results. That did not halt the search.

The ranks of researchers attempting to explain why X-rays are not visible to the human eye, normal or blind, grew with the years. The relative impermeability of the lens to X-rays was the reason commonly given. Some concluded on further testing that the rays were not at all visible even to the blind, that there was no variation to explain. A few, however, following Brandes and the later Roentgen, found evidence that the rays are seen by many or most eyes, possibly due to fluorescence of the eye matter or sensitization of the retina on exposure.

D.D. Bossalino reviewed many of these findings (not Foveau's) and performed his own series of tests to determine whether X-rays could be seen.[36] He wrapped his eyes with a secure blindfold that did not admit light, and positioned himself facing a Crookes tube powered by a moderately powerful coil at 25 centimeters distance. When the tube was switched on he perceived a clear greenish opalescence but none of the objects in the room around him. A sheet of cardboard or of wood did not obstruct the light, but a lead screen did. He was able to discern the number of copper wires held between his blindfolded eyes and the tube, and made out any arrangement of the wires in relation to each other.

When Bossalino placed a thick lead plate before one eye and a metal sheet crossed by two indentations perpendicular to each other before the other he saw a cross of luminous bands that glided left or right depending on how he moved the plate. A sheet crossed by indentations of varying size presented a shining greenish disc sized in proportion to the diameter of the line most in view. The X-ray light formed geometric projections of the indentations of the metal sheets in the fluorescence of the retina like shadows in intense light.

---

[36] Bossalino (1906)

Experiments with other normally sighted people and with people whose sight was impaired by cataracts brought the same results.

Bossalino concurred with Roentgen and a few others on the ability of X-rays to project metal shapes by means of the fluorescence they induced in the eye that was not receiving light rays and was adapted to the dark. The fluorescence was the light that the retina received. The metal shapes were shadows in the fluorescence. The crystalline lens was no obstacle to the rays whether clear or clouded by cataracts. These conclusions were summarized in European and American medical journals. Given the comprehensive nature of Bossalino's study it seemed settled that X-rays could in themselves be seen by anyone with a functioning retina.

Bossalino did not experiment on the blind. The fluorescence mechanism for seeing X-rays precluded any belief that X-rays could impress themselves or the image of any object directly upon the optic nerve and brain. The blind were not likely to see through the medium of X-rays. In the following years X-rays like other forms of radiation were used therapeutically for specific eye conditions.[37] The deliberate aiming of X-rays into the eyes of the sighted or the blind no longer was an active research method, however many other X-ray projects of dubious safety were undertaken.

There was no constant source of what X-rays looked like to feed into a culture cognizant of the uses of X-rays to penetrate bodies, stressed metal components and sealed containers. The entertainment fluoroscope and the X-ray photograph of the hand were known in the cities, but they were a limited demonstration of what eyes that could look into anything must actually see.

---

[37] Law (1934)

4

PUBLIC X-RAYS

Seeing inescapably with X-rays displaced the fears and regrets of colonialism amid changing domestic routines. The generally amused spirit of receiving X-rays into public life covered these fears. Historians of X-ray acceptance have recognized the humor and the anxiety beneath it. People in the early twentieth century consciously adopted a mock serious tone when responding to the revealing rays.

A paragraph at the bottom of the page in an 1896 issue of a trade journal, *Electrical Engineer*, reports a legislator's initiative.

> X-Rays in Opera Glasses
> A loud laugh went over the State of New Jersey on Feb. 19 when Assemblyman Reed, of Somerset County, introduced a bill in the House, at Trenton, prohibiting the use of X-Rays in opera glasses, in theatres.[38]

The Assemblyman's proposal is often mentioned as an example of the ignorant rush to counter imaginary dangers of X-rays. From reportage of the time it appears to have been a joke which everyone, including the Assemblyman, shared.

Opera glasses were ornate binoculars theatre audience members used to scan each other to pick out familiar faces and detect gossipworthy fashion and liaisons. Adding X-rays was an imagined extension of their existing capabilities. The suggestion of government intervention to prohibit opera glass X-rays brought to the fore the

---

[38] *Electrical Engineer* 21, 408 (1896): 216

possible violation of public order as the social surveillance entered clothing.

One of the letters Thomas Edison later made public was a request that he fit a pair of opera glasses that accompanied the letter with X-ray lenses.[39] This was treated as another absurd request of the great inventor. Another letter writer sought Edison's assistance in constructing glasses that would enable him to look into the cards held by other players at a game of faro.[40] Any faro player would know that no experienced player would take part in a game with a player wearing unusual glasses. The spirit of X-ray play covered these proposed inventions as well.

Edison both encouraged and discouraged the belief that X-ray operated visual equipment was in the works. His May, 1896 fluoroscope display was soon followed by an exhibit at the American Museum of Natural History of a pair of X-ray spectacles.[41] His industrial research facility had evaluated hundreds of fluorescent compounds and determined that tungstate of calcium was most brightly illuminated by X-rays. A surgeon wearing spectacles fitted with calcium tungstate-coated lenses could view the bones of a limb placed between the spectacles and an X-ray source. Surgical X-ray spectacles were advertised by different manufacturers for decades.

Edison's X-ray spectacles weren't needed to encourage the spread of humorous visions of X-ray vision. These visions had two aspects: an individual wearing the legendary spectacles gazing at a person we are to understand is opened up physically and mentally, and a general view of the world as X-rayed. The zahori and the phantasmagoria views are still present. The former led to a routine but culpable subjectivity of X-rays that was the most persistent form embodied in X-ray spectacles. The latter was subsumed into satirical visions before giving way to the cinematic phantasmagoria of reanimated skeletons and cutaway houses that didn't require an X-ray explanation.

---

[39] Glasser (1993: 203)

[40] Jones (1907: 263-64)

[41] Things New in Science, *Richmond Dispatch*, March 29,1896: 3; X-Ray Spectacles Now, *Guy's Hospital Gazette* 10 (1896): 187

An early 1896 illustrated column by the cartoonist Walt McDougall explores the likely consequences of the predicted adoption of X-ray spectacles within the following year.[42] A gentleman will look into his epiglottis and see the adhesions from the previous night at his fraternal lodge; or into his appendix to see the residue of a rich meal; or into the pocket of an acquaintance he plans to "touch" for five dollars. A housekeeper can detect the chick embryo in a "strictly fresh egg" she is about to buy. A wife can dissolve a thick oak door on the other side of which her husband sits drinking a glass of (of course!) mineral water. And a gambler can see what Edison's faro-playing correspondent hoped to see.

The one X-ray spectacles illustration McDougall made for this column doesn't refer to any of these in particular.

CATHODE SPECTACLES IN THE HOME.

6. Cathode Spectacles in the Home. Fn42

---

[42] M'Dougall Moralizes-Stray Thoughts on Science and an English Butler with Roentgen Glasses, *Roanoke Daily Times*, March 28, 1896: 6

The wife wearing her cathode spectacles as a lorgnette gazes directly into the skull of her flummoxed husband, reading his thoughts.

The rest of McDougall's piece focuses on the other invasion of X-ray spectacles, into clothing. McDougall sees this as burlesque moral revolution.

> One feature of the so-called cathode ray discovery is the certain fact that it will restore mankind to a state of natural simplicity. Prudishness will be completely eradicated and a Hellenic condition of dress will prevail, for it will cause us to dispense entirely with clothes-at least in the summer months. Antinudity and living picture edicts will not be enforced for in that day our forms will be visible in Trilbylike entirety. No one will bother about clothes that do not conceal his perfect figure from others wearing these improved eyeglasses, and of course all will wear them, for none will wish to be at a disadvantage.

McDougall envisions the transition from the individual domestic X-ray spectacles to the revolutionary universal transparency, the opera-glass condition of the entire world, where none will wish to be at a disadvantage.

Governments may have been jokingly called upon to halt the use of X-ray glasses. In Europe and America they had banned theatrical ("living pictures") and photographic displays of the unclad human body, always making allowances for the Hellenic, classical Greek sculptural or athletic body. Trilby O'Ferrall was a poor girl working as an artists' model in the fin de siècle Paris of George du Maurier's 1895 novel. McDougall refers to the condition in which she posed for the enraptured painters. Esthetic nudity was making its entrance into bourgeois society. X-ray eyeglasses were one more way to allude to that nudity without actually depicting its nakedness.

X-ray technology was quickly assimilated to looking and being seen within society. The X-ray eyeglasses were a surrogate for the eyes themselves always searching for novelties and disorders. Fears of surveillance and mind reading anticipated actual surveillance reconstructing the everyday gossipy attentiveness of village life, ever preoccupied with social status signs, within the anonymity of urban life. McDougall was a master illustrator of the mass scene where social restrictions are unseated, an anticipation of the slap-stick cinematic comedy to follow. McDougall also was one of the illustrators of the Oz books, and among the comic strips he drew was one based on the Oz characters.

A cartoon William Glasser reproduces in his study of the early response to the Roentgen rays is a panel from a Paris periodical, *La Nature*, Variations sur les Rayons X.[43] No one is wearing X-ray spectacles, which seems at first to have been an exclusively American response to the advance of the rays. A multi-story house is in cross-section with the activities of the inhabitants all in full view; the contents of a purse are revealed, as is a skeletal arm wrapped around a portmanteau. A man hovers over a sleeping woman, a cathode ray tube aimed at her head (her thoughts known to him now) and a customs agent inspects a large chest with the help of the rays. A carte de visite frame surrounds the drawing of a skeleton surrounded by a man's clothes; the edge of a woman skeleton's card is visible beneath that one. And below, males and females dressed in metal plate stroll in the street. The probing X-rays will not gain access to the depths of their clothes.

This series portrays both the public exposure and the individual-looking mode of the X-rays. X-ray cartes de visite were produced in Europe and America, though the full-body X-rays of the drawing are a fantasy. Being stripped bare on the street by the broadcast X-rays that can section a house was a threat to arm against.

Almost every commentary on the introduction of X-rays repeats the canard that within the first month after X-rays were made known a "London firm" advertised "X-ray proof underwear." The underpinnings of this rumor are apparent from a brief notice that appeared in the weekly technical newsletter *Electrical World* on

---

[43] Glasser (1993: 40)

March 28, 1896, at the very end of an issue with a great deal of information on X-ray technology.

> Commercial Application of "X-Rays"
> We learn that in London, where honesty is not al-
> ways considered to be the best business policy, a
> firm is making prey of ignorant women by advertis-
> ing the sale of "X-ray proof underclothing."[44]

The only advertisement for this underclothing that anyone has been able to unearth is a cartoon of a man tugging at the stays of a wom-an's cramped bodice.[45] There is no evidence that X-ray proof un-derclothing ever was purchased, whether by "ignorant women" or curiosity collectors. Both the X-ray opera glasses and the response to the possibility they might exist were tones of the same laugh.

Commentators on the sights before the eyes of X-ray spectacle users took advantage of the opportunity, as in the *La Nature* cartoon, to bait hotel guests and ladies out in public. A column in a newslet-ter for phonograph hobbyists gave an almost cinematic account of the walls between rooms melting away when someone turns on his pocket battery while wearing X-ray spectacles. Ladies, the author continues, might consider adopting a lightweight, durable alumi-num bodice and skirt.[46]

> Some such protection would be indispensable, for
> it is easily to be imagined how horrifying it would
> be to a lady on the street to find out she was being
> ogled by an X-ray fiend.

Ladies might consider any ogling in the street horrifying. The X-ray element technologizes the fright.

---

[44] Commercial Application of "X-Rays", *Electrical World* 27,15,March 28, 1896: 340

[45] Henriksen and Maillie (2003: 6)

[46] Glasser (1993:46)

The phrase "X-ray fiend" seems to have been a passing expression for just this type of ogler, who might just as well have been equipped with X-ray imaging. It was one of the titles given to a brief trick film of 1897. George Albert Smith, who already had filmed a mischievous ghost bedeviling prop movers, spent 48 seconds to bring a man bearing camera-shaped box labeled "X-RAYS" to a park bench where a man and woman are engaged in theatrically allowable fondling. The cameraman removes the cap of the lens pointed at the pair and they continue their play instantaneously transformed into skeletons (or people wearing black outfits with human bone structure painted on). With the lens cap back on they are fully clothed again, and the film ends.

7. Still from The X-Ray Fiend (1897), George Albert Smith

Apart from containing one of the first cinematic kisses (and between skeletons no less) Smith's film *The X-Ray Fiend* includes the skeletal umbrella and a measure of the anxiety about X-rays, or about film itself, penetrating and recording intimate moments without the knowledge or consent of the X-rayed.

The cinema at its beginnings turns the camera itself into a metaphor of the X-ray machine removing the veil of clothing and skin from the amorous pair. It irradiates them rather than recording their activities. By making them skeletons it removes the possibility of objection to fleshly contact while engaging in the game of eavesdropping upon lovers that had long been a feature of novels and photographs.

Georges Méliès soon made his own minute-long contribution to the growing cinematic engagement with X-rays, *Les Rayons Roentgen*, released in England as *A Novice at X-Rays*. The French magician screened a man visiting a doctor to demand an X-ray. The doctor activates the machinery and the man's skeleton leaves his envelope as skin and organs fall to the floor. The doctor reverses the process, the man refuses to pay, and the machine explodes. The elements of realism, the demand for an X-ray, the assumption that it would reveal what is wrong inside the man, and the refusal to pay when expectations aren't met, counterbalance the fantasy of skin slippage and machine explosion.

This film is not unlike the many other appearances of skeletons in early films, including those of Méliès. They all are part of the animated skeleton tradition inherited from the phantasmagoria. *The X-Ray Fiend* and *A Novice at X-Rays* illustrate the transition from fleshed body to skeleton by means of an X-ray device made visible to the film audience. The X-ray machine broadcasts X-rays over entire figures but not the entire screen.

The film camera includes the X-ray image in its broader panorama, just as photographs were imagined showing both clothes and skeleton within them (but never the skin). Representing the X-ray screen photographically frees illustrators and cartoonists from having to intrude upon the skin beneath the clothes. The covering layer simply falls away or is dissolved. The two films epitomize and anticipate the practice of public X-rays by means of tubes, cameras and spectacles. What is expected to be seen is implied by what is shown.

An Art Deco framed magazine drawing from the first year of X-rays unfolds a fully irradiated scene.[47]

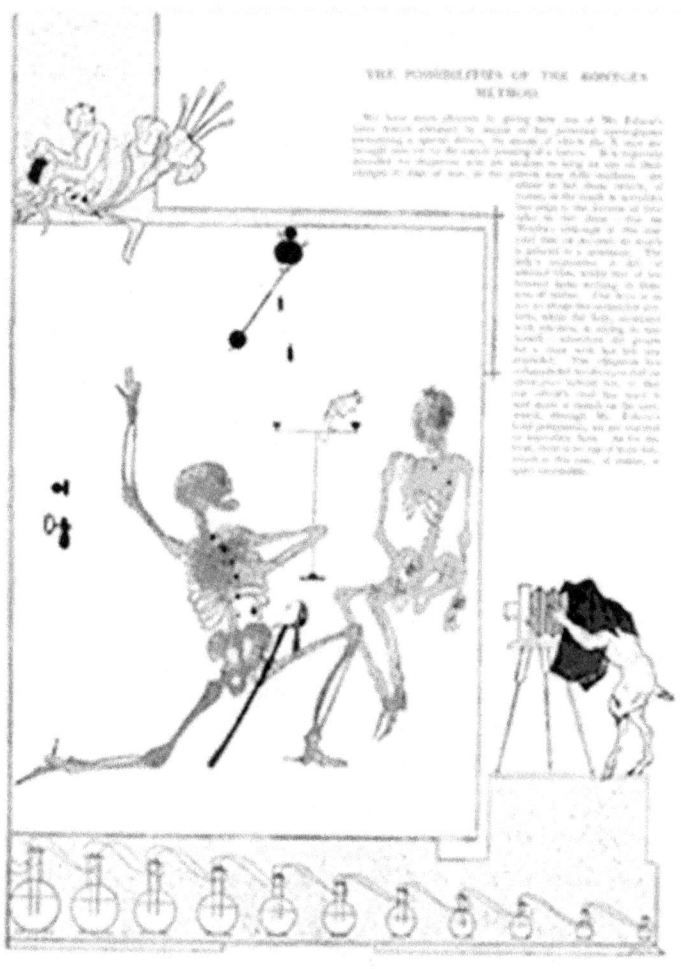

8. The Possibilities of the Roentgen Method. Fn47

---

[47] The Possibilities of the Roentgen Method, *The Picture Magazine* 7, January-June,1896: 253. This illustration was reprinted in the German magazine *Dur und Moll* as "Die Liebeserklärung" ("The Declaration of Love") in 1900.

The satyr-operated camera and illumination from above have removed clothes and skin from the military officer making a vow to his seated lady. Doorknob and lock, clock pendulum and weights, the cavalier's sword, the row of brass buttons down the front of his uniform, his spurs and the lady's brooch and heart jewel are all in shadow relief. A skeletal parrot occupies his perch. A row of batteries connected in descending series, the power supply for a Ruhmkorff coil, forms the bottom border.

The text's story of the scene differs from what is shown. The officer's rival has gazed through the intervening door with the chaperone's Edison-Roentgen opera glasses and made this sketch of the courtship. Both modes of the X-ray panorama, broadcast irradiation and X-ray spectacles, are therefore enfolded into the same comic drawing.

The Picture Magazine where this drawing was printed was the pictorial companion of *The Strand*, one of the chief English literary and cultural magazines of the late nineteenth century. The magazine included a range of photographs, prints and drawings, informative, sentimental, comic. "The Possibilities of the Roentgen Method" had been the title of a collection of X-ray photographs illustrating medical, investigative and commercial uses of the rays previously published in *The Picture Magazine*. The present drawing was a satirical reflection of one more possibility of the method. Two more, if you include the chaperone's possession of the X-ray opera glasses to keep track of her charge's doings behind closed doors.

Soon after the discovery of X-rays a public space was opened up where anyone might see the surrounding world as an X-ray photograph or fluoroscope rendition of the whole. Either X-ray glasses or a device that poured exposing rays over the entire scene was the technical means. Representation of the results was limited, conveyed by hints and both blocked and further represented by defensive garments, and stopping only at the skeleton. The X-rays mingled with another visual novelty, the cinema.

This world of reaching and inhibited sights did not originate with X-rays. Being publicized as a way of looking into the body with perhaps a stop at the skin-clothing boundary gave the rays, or their technical means, an attachment to existing prospects and anxieties.

The historical and ethnographic record contains instances of notice-able enhancements of the eyes being interpreted as invasive sight, perhaps activating evil eye beliefs with respect to that person.

Lt. Col. William H. Emory, halting his military reconnaissance expedition of the American southwest for November 13-14, 1846, observed the reaction of some Maricopa Indian women to the eye-glasses worn by a civilian member of his party.[48]

November 13 and 14.—With the morning came the Maricopas women, dressed like the Pimos. They are somewhat taller, and one peculiarity struck me forcibly, that while the men had aquiline noses, those of the women were *retroussés*. Finding the trade in meal had ceased, they collected in squads about the different fires, and made the air ring with their jokes and merry peals of laughter. Mr. Bestor's spectacles were a great source of merriment. Some of them formed the idea that with their aid, he could see through their cotton blankets. They would shrink and hide behind each other at his approach. At length, I placed the spectacles on the nose of an old woman, who became acquainted with their use and explained it to the others.

Norman Bestor was listed in the expedition roster as an "assistant". Emory recounts several occasions on which he preserved valuable chronometers and other instruments from ruffians and natural disaster. Responsible for taking readings, he seems to have been the only member of the expedition wearing such conspicuous eye circles.

The Maricopa women didn't treat the novelty with fear, but as a source of play, similar in spirit to the reaction to the imaginary X-ray opera glasses in the east fifty years later, down to the fun-filled urge for protective barriers. Emory, a matter-of-fact narrator, doesn't say what the effect was of the old woman explaining the glasses to the others. I imagine she said that they caused a distortion of vision. They probably had some fun with that too (Why would someone wear something that makes everything go blurry?).

During the 1960s revival of X-ray spectacles the spirit of fun and invasion continued, and a distinct form of distorted vision was sub-stituted for X-ray photographic sight. This first period of X-ray spectacles was a frisson of social seeing through a surprising new technology. X-ray spectacles are seldom mentioned in popular or

---

[48] Emory (1848: 87)

scientific media after 1906 until the 1950s. They were relegated to a children's whim.[49] Or removed to a technological Olympus.[50] The X-ray sight metaphor was beginning to replace the optimistic certainty that technology would soon develop a means for everyone to see with X-rays, as it had made it possible to see permanently and speak remotely.

Imagining the sight of the passing crowd through X-ray spectacles was imagining private nudity made public and freely available. The institutions for procuring nudity were socially marginal and subject to interdiction by a range of legal and moral authorities. X-ray spectacles or projectors made a world in which these boundaries were rolled back in secret or publicly by an outpouring of rays. As Crosthwaite and some spectacle-makers concluded, random exposure of the body would be uncontrollable and disconcerting. Stage shows that used X-rays or other rays as the excuse for displaying unclad bodies included a skeletal stage after the skin, as the momento mori part of the titillation.[51] The comedy of X-ray revelation was a black comedy verging on horror formed of the fear of death that had accompanied X-rays from their first demonstration, and which the comedy seems to be made to displace.

Anyone wanting to see what bodies would look like through X-ray spectacles had only to turn to the medical journals, which added (when technically feasible) X-ray exposures of the raw human flesh it was the business of doctors to examine and report to their colleagues. Even if a view of those pages was possible for someone seeking an X-ray spectacle, the strictly clinical depictions held little to excite most would-be viewers.

---

[49] For example Biddle (1902: 226-27) relates the naïve belief of two girls that a phrenologist's eyeglasses are X-ray spectacles that allow him to see into the heads of his subjects.

[50] John Kendrick Bangs (1902: 124-25) on a visit to Olympus is diagnosed by an X-ray spectacle-wearing Aesculapius, who assures him that his children, the physicians, have the glasses but don't really need them.

[51] Pang (2007: 199-200) describes a Shanghai backroom show of the 1930s that used "YS rays" for this purpose.

Internists and orthopedists in Europe and America during the early twentieth century used X-rays to study the effects of tight corseting on the spine and internal organs. Achieving an hourglass figure had consequences most visible to the specialist in the part of the body X-ray underwear was meant to protect from the prying rays.[52]

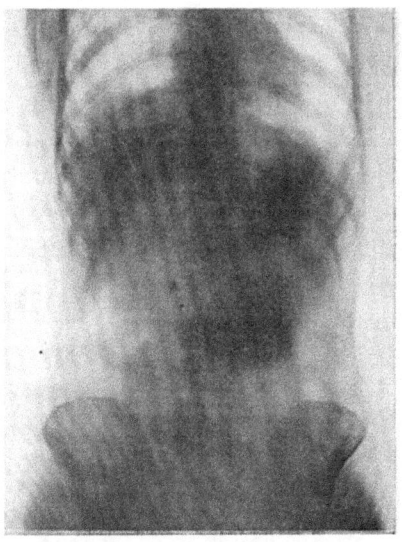

9. X-ray photograph of a woman's corseted torso. Fn51

Changes in fashion, and perhaps these gastrointestinal discomforts, led to the relaxation of corseting. By the late 1930s a corset manufacturer was offering potential customers X-ray photographs of internal organs compressed by rivals' products, and relaxed within their own superior sheathing.[53] This advertising encapsulated the history of figure-shaping undergarments in X-rays, and the availability of visions thereof.

---

[52] Lion (1906)
[53] Pictures Rarely Seen Outside a Doctor's Office, *Life*, August 12, 1940: 4

The disappointment of donning X-ray spectacles and seeing the effects of corseting on the desired body is not suffused with horror. An application of X-rays in the eyes research comes closer.

Thomas Edison held that X-rays, more penetrating than light, should break through the barriers to sight and enable the blind to see. His widely publicized experiments with two blind men supported this surmise to the extent of bestowing on the two men a lasting capacity to discriminate shapes and motion. More work was needed, Edison said.

Alexander van der Naillen, for one, was not surprised at Edison's findings. Van der Naillen was the founder of engineering schools, the longest lasting one in Oakland, California, and a prolific author of theosophical treatises and travelogue-novels of his adventures gathering ancient wisdom from the lamas of high Himalay. Asked by a reporter for a San Francisco newspaper to comment on Edison's X-ray treatment for blindness, he shared the great hopes of his contemporaries.[54]

"X rays are going to develop marvelous discoveries soon. My son Ralph handles the fluoroscope in the laboratory, and the rays have had a wonderful effect upon him. His hairs have fallen out from the backs of his hands and the nails have dropped off his fingers through constant use of the instrument during two months. This shows the intense physiological influence the rays have upon the human body. The rays penetrate the innermost parts of the body. Thus it behooves doctors to take the most earnest interest in those rays. To the physician's domain belongs a thorough investigation of those mysteries, because it is no longer a matter of mere curiosity, but one of deep scientific concern.

"If the rays can kill that capillary cell in which the hair is rooted, why would it not be worth while to investigate their action upon those abnormal growths, such as cancer, tuberculosis and kindred diseases?

---

[54] Can the X-Ray Restore Sight? *San Francisco Call*, November 19, 1896: 7

After a revelation he received in the Napa Mountains, van der Naillen remained certain and paternal about X-rays: "and still these rays are the children, the offspring of electricity, the legitimate grandchildren of magnetism."[55] His own son apparently survived to travel (or flee) with his brother to Yucatan, where they procured more ancient wisdom from the Maya.

Shot directly into the eyes, the rays certainly did have an intense physiological effect on the human body. The shift from genuine experiments to the plane of spirit is, however, understandable.

An alternative pathway that did not evoke X-rays opened in the envisionment of the human aura, "a force emanating from the human body," through glass lenses formed of plates with a layer of dye solution between them.[56] This concept of a spiritual technology merged with the X-ray spectacles to produce hybrid fictions.

Comic author Noel Godber's novel *Amazing Spectacles!* follows Freddie Wynne and his comrades who commercialize a deceased uncle's legacy of a chemical powder that dissolved in water placed inside the glass of a lens enables the wearer of spectacles or user of the camera to see through clothing, doors and other barriers.[57] Social disorder results, no secrets are safe, men wearing spectacles are mobbed by women, a politician is embarrassed, and "antiseptic" doors are advertised. In the final chapter Freddie and partners are preparing to merchandise a newly invented technique for imbuing cloth with metal, rendering it impenetrable to the prying chemical vision as they are to X-rays.

Godber's novel is a continuation of the decades-old public X-ray environment in its comic and uneasy dimensions with the assistance of an old spiritualist technology drained of its metaphysics. At the time the original X-ray vision was being revived under new terms, in what would become its definitive form.

---

[55] Van der Naillen (1912: 43)

[56] Kilner (1911)

[57] Godber (1931). The *Futurian War Digest*, 1, 12, September, 1941: 3 noted the reprint of Godber's novel in the Jolly Novels series, and did not recommend it to readers.

5

XYLOPE AND ANNA B.

"Who hasn't tried to represent to himself the worldview of someone constructed to perceive these rays directly-for which it would suffice for him to have an eye of wood or cardboard?"[58] M. Gaston Moch inquired at the beginning of an essay on the relativity of human consciousness.

Moch was intrigued by William Crookes' January, 1897 address before the Society for Psychical Research, "The Relativity of Human Knowledge," which had been translated into French as "De la relativité des connaissances humaines."[59] The British scientist, with a bow to Jonathan Swift, postulated a "homunculus" positioned on a cabbage leaf perceiving and subjected to the forces of his station invisible to us with a larger view. Separating the grades of vibrations that would include all forces and matter, Crookes expanded the relativity of the mathematician William Hamilton to include more recently discovered X-rays and radioactivity. Crookes contented himself with his cabbage-leaf homunculus amid the temporal relativity of William James.

Moch's essay was published in the same journal as the Crookes translation three months later. He adopted Crookes' method of describing the experience of a being at another grade of the physical world. His X-ray-perceiving relative consciousness was a being with wooden eyes, named in a Greek compound "*Xylope*." The Xylope sees only in rays that can pass through solid matter, such as wood. We see with light in vibrations, 450-750 trillion per second; the

---

[58] Moch (1897: 104)
[59] Crookes (1897)

Xylope sees with invisible radiation 300-2300 quadrillion per second. What would he see relative to what we see?

> Of his beloved, he will only perceive the skeleton, surrounded by a confused and translucent mass presenting to our sight an aspect we consider gelatinous! The criteria of beauty for him will not reside in the details we can admire such as an expressive eye, a well-shaped mouth, white teeth...(or black teeth, if we happen to be Chinese). What one will read in a novel, will be this sort of description: "Ernestine is endowed with a thoracic cavity of impeccable symmetry, bounded by two discs (omoplates) of the purest design; her graceful sternum and above all her marvels of position (cubitus), in contours delicately marked by the semitransparency of the flesh, and so on.

The Xylope hides himself away in a house of glass walls (glass Moch believed opaque to x-rays) and receives the radiation of day through wooden shutters after the glass windows are opened. The Xylope sees the forest as a collection of fixed fountains of sap rising with the spring and stilled during the winter. As his civilization advances he learns to remove these fountains and using tools shape them for his use, all the while avoiding the pain of colliding with the invisible woody mass that surrounds them.

This is the limit of Moch's excursion into the world of beings who can see in X-rays. Other kinds of seers might see the traces of light which has already vanished, having a sight of events frozen in time or scrolled backward to the beginning. The fall of a man from the heights of the Eiffel Tower might be halted or reversed. Some might see a world smaller than the cells that are considered the lower limit of life, in two dimensions or even one dimension.

Moch's essay reflects shifts in size perspective contemplated during earlier periods when the microscope and telescope were new, by authors such as Voltaire and Swift, and more recent fictions of altered geometry, such as Abbott's *Flatland*. The controllable time of cinema also is an influence. His chief observation is that there are many different perspectives possible each of which seems to be the dominant one to those who see the world that way. These perspectives can only be explored by understanding they are relative to our own equally enclosed way of looking at things.

Moch's projection of the Xylope is a method for jolting his readers out of their accustomed views, and preparing them to accept other physical viewpoints. The Xylope point of view was not an end in itself. It opens other views of the world. Like his predecessors and contemporaries who placed their voyagers in a world of changing size and lineaments, Moch was a satirist. He envisions the conventions of the romantic novel dissolved into an anatomy text, and places his Xylopes in a forest of wonders, dangers and possible enterprise. His allusion to the alleged Chinese preference for cosmetically blackened teeth is in a spirit of cultural relativism to correspond to the physical relativism he is promoting.

The Xylopes are a possible evolution, a stage along the harmonic scale of sensitivity to vibrations that constituted human potential for scientists like Crookes and Moch. The Xylopes are a thought experiment like Einstein's fixed and moving observers relative to each other, but within a psychological and cognitive framework rather than Einstein's cosmic geometry.

Moch was the son of Col. Jules Moch, one of the first Jewish officers to advance into the upper ranks of the French army. In the steps of his father, Gaston Moch graduated from the École Polytechnique where he was a classmate of Alfred Dreyfus, whom he would openly defend against the charges anti-Semitic factions used to secure Dreyfus' arrest, trial and imprisonment.[60] An artillery captain, Moch wrote pamphlets on advanced weaponry while recoiling from the prospect of a war waged with those forces. He reflected back upon the Franco-Prussian war and the American Civil War, and forward to the technology of future conflicts and the dangers of the heavy armaments introduced to the border between

---

[60] Bourrelier (2008) details Moch's Dreyfusard and pacifist careers but does not mention his relativity moment.

France and the victorious German empire that had swallowed Alsace-Lorraine.

Moch was a practical internationalist who attended peace conferences, advocated aboriginal rights in the colonial possessions, and sought an international language to ease communications across national borders. He was fluent in French, German and English, and he wholeheartedly embraced Esperanto, becoming executive secretary of the Esperanto office in Paris. His translations of a French-English grammar and French literary works into Esperanto are among his other writings in a list found at the front of a 1922 publication. It doesn't include the narrative of his balloon adventures over the Sahara desert.

10. Portrait of Moch by Felix Valloton, from *La Revue Blanche*

Moch did not recall his Xylope excursion into the relativity of consciousness until he wrote an introduction to Einstein's theories after the First World War, one of many books on this subject, as he noted in the preface. He recalled the early instance of "amusing himself" with changing perspectives.[61] He returned to the Voltaire-an contrast of ant-like and giant points of view, and introduced the work of a colleague who inferred painters' eye problems from their styles. The following year, reprinting his Xylope essay as part of another book, he placidly confessed that his relativity "very certain-ly was far from holding the seed of Einstein's theory."[62] At the same time he gave the social theorist Gustave Le Bon and the feminist-pacifist Clémence Royer credit for anticipating Einstein's philoso-phy, which did not endear him to fellow Esperantist Einstein and his close supporters.

Crosthwaite's X-ray eyedrops and the removable X-ray specta-cles were a temporary condition that passed as soon as it had served its purpose. Moch's Xylope occupied a rung on a scale of vibrations and knew no other world, giving his author the freedom to imagine what would constitute beauty for such a creature, and how he might utilize natural resources visible in a very different form. Neither Swift's Gulliver nor Voltaire's Micromegas but Edwin Abbott's "A Square" of *Flatland* (1884) is the closest literary analogue, though "A Square" also sought to escape his two-dimensional space. Crookes, James and Einstein all employed relative observers and like Moch quickly exhausted the descriptive scope of any one stance.

Moch himself revised the Xylope when his article was repub-lished the following year in an astronomy journal.[63] In addition to Crookes' relativity address the beginning of the article now refers to the inspiration of astronomer Camille Flammarion, the doyen of the journal, and makes his invisible concurrent world into an inter-planetary one. The same discussion of size and time relative observ-ers precedes the Xylope, but now with obeisances to Flammarion's interplanetary speculations. This turn to the extraterrestrial Xylope influenced other contemporary writers, who translated or para-phrased the Xylope sequence in their own articles.

Ramiro Blanco titled his translation of the Xylope text into Spanish, "El Mundo de los Rayos X" ("The World of the X-Rays"),

---

[61] Moch (1922: 12-13)

[62] Moch (1923: 135)

[63] Moch (1898)

in which he situates the wooden-eyed X-ray seers on another plan-et.[64] The Abbé Moreux, writing of the Xylopes both in a Catholic spiritist journal and in an astronomy journal,[65] links Voltaire's in-terplanetary Micromegas to Moch's creatures. Xylopes on other planets were confined to the Continental commentaries on Moch.

Moch's article was widely noticed in English and American journals and newspapers. This paragraph from the English trade newsletter *Pharmaceutical Journal* charges Moch with making a judg-ment that his relativism precluded.[66]

> XYLOPES ARE THEORETICAL INDIVIDUALS, whose existence in the future is predicted by M. Gaston Moch in a paper in the *Revue Scientifique.* He assumes the possibility of persons being specially gifted with the faculty of seeing countless vibrations of rays of light, and seems to think that the development of the application of X rays will lead to the discovery of many beings gifted with this peculiarly piercing sight. To such persons, he observes, the most beautiful human countenances would appear repulsive, as they would appear as skeletons covered with a sort of gelatinous matter. Needless to say, M. Moch expresses pity for the hypothetical persons whom he calls xylopes, and he appears to ignore the fact that if such beings as he imagines were to exist their standard of beauty would be based on their own experience, and not on the impressions of individuals gifted only with average sight.

By confining his comment to the relative assessment of beauty the writer does not acknowledge the scope of Moch's confabulation.

The English-language commentators all seem to have taken their knowledge of the article from a summary that excluded the Xylope houses and adaptation to forestry. How unpleasant it would be to see the world, and especially women, in this translucent way.

The American journalist Leroy Mosher, writing one of the earli-est English-language pieces on the Xylope,[67] was especially taken with, and repulsed by, the gelatinous mass enveloping the skeleton

---

[64] Blanco (1900)

[65] Moreux (1902-03); Moreux (1905)

[66] *Pharmaceutical Journal* July 31, 1897: 93

[67] Mosher (1905: 346-47). Mosher's piece was printed in the *Los Angeles Times* on October 2,1897.

of the beloved in view. Mosher was certain that human nerves are too weak to stand the strain of seeing the "ghastly promenade of rattling bones" in the streets of populous cities. The world has "worried along" for thousands of years without the Xylope, and can just as well get along without him. He does not "fill a long-felt want."

Mosher is in accord with Crosthwaite's Herbert Newton, who takes to his bed on seeing just one woman in that state, and with the author of an editorial paragraph that appeared in a number of American newspapers in late 1897.[68]

We May All Have X Ray Eyes.

The news comes from France that we are eventually to have X ray eyes. An exceptionally learned French scientist says that in time we will be enabled to see almost any number of vibrations of light.

Not all of us may be so gifted, however. The faculty will be confined to a few, and they will be called xylopes, if the French professor has his way about it. To these xylopes the French scientist says that lovely women will appear as skeletons covered with a gelatinous sort of matter.

What value, then, will the human form have! There will be no roses, no dimples, no pretty curves. Grace will be typified by bones—ugly, horrid bones. Laughter will be ossified, tears will be invisible, weeping a mere rattling of teeth. The streets would be filled with the horror of lipless, hairless lidless crowds. The terror of pervious clothes would vanish. Modesty would assume new forms and phases. Or perhaps we would go back to medieval days —to the suits of armor—and go about the streets clanking and unhappy.

The Xylope's vision certainly violates the conventions of the (female) human form that prevailed at the turn of nineteenth century. The sentimentalism in conflict with realism on exhibit in all newspapers is epitomized in the article. Moch's revised sentimental novel has been turned into the metallic outerwear at the end of American happiness.

A motif that unites Moch with all of his commentators, which was defused by the extraplanetary version of the Xylope, was that sight on other vibrational levels is a coming result of evolution. No

---

[68] Example from *Kansas City Journal* December 5, 1897: 5

one brings up the subject of how seeing with X-rays would confer a selective advantage. It's an outcome we would prefer to avoid. Being forced to look at things best kept out of sight is the direction of our present evolution, and we don't like it. "We May All Have X-Ray Eyes" and can't help it.

After this brief spurt of attention the Xylope was forgotten, but of course, X-rays were not. They were among the abilities cultivated by members of the political left as they faced the future.

Before X-rays were announced, Anna B. was acquiring vision appropriate to them. She was said to be plain, pleasant but unprepossessing, a working class woman of Narbonne, France, which is precisely the reason she came to the attention of Dr. Ernest Ferroul in 1894.

At that time Dr. Ferroul was mayor of Narbonne, having been elected in 1891, and held that position with only one three-year break, through strikes and arrests, through efforts by the French state to suppress his attempts to foster a socialist government in the Languedoc city, until his death in 1921.[69] The son of artisans, Ferroul received his physician's training at the ancient medical school of Montpellier, and establishing his practice in Narbonne, soon became known as "the physician to the poor."

Anna B. came to him as a patient, subject to fainting spells and bouts of confusion. Ferroul's hypnotism treatments improved her state of health, and made him aware of her ability to narrate events taking place at a distance as they were taking place. After conducting several experiments to confirm his observations, Ferroul brought Anna B.'s seeming clairvoyance to the attention of a Montpellier professor of clinical medicine, Dr. Joseph Grasset.

Grasset already was the author of several texts on the treatment of nervous disorders, and was shifting his attention to the rich field for study and experiment in neurasthenics with apparent second sight. Grasset approached Anna B. with scientific detachment, hesitant to appear over-eager to lend his professional weight to the assertions of his less detached colleague. An engineer, M.A. Goupil, became the medium of transmitting the Anna B. events to journals

---

[69] Guthrie (2010)

in the form of his own accounts and correspondence between Ferroul and Grasset.[70]

Public exposure of Anna B.'s hitherto circumscribed reputation seems to have been driven by one letter sent by Grasset in Montpellier to Ferroul in Narbonne.[71] It consisted of two lines of verse, and words in Russian, German and Greek on a piece of paper sealed in an envelope which was enclosed in metal foil used to wrap chocolates in such a way as to preclude any tampering.

Anna recited the verses when shown the sealed letter, traced the larger Russian word (the name of the writer and physician Chekov) with her finger and said she could not discern the other words, except for the city and date.

This triumph precipitated her downfall, which coincided neatly with Ferroul's years out of office, and Grasset's pullback from involvement. An investigation by a panel of Narbonne's Academy of Science and Letters cast doubts upon the authenticity of Anna's readings. After not many years she was relegated to a footnote in the history of clairvoyance where the word "fraud" also appeared.[72]

In his comments on Anna's successful reading Grasset did give her the means of a little additional life.

---

[70] Goupil (1897)
[71] Grasset (1897)
[72] Podmore (1908: 340fn1)

> This is truly reading through an opaque body, tak-
> ing the word "opaque" not only in its ancient sense,
> but also in the scientific sense given it by the dis-
> covery of the X-rays.

Opaque did not just mean blocking light rays and therefore vision,
but an entire suite of other rays which were or were not passing
through matter. Grasset did not mean that Anna had X-ray vision,
but rather that she was able to see with unknown rays that could
pass through metal opaque even to X-rays. In the wonderment of
the moment, Grasset envisioned undiscovered rays which the Narbonne woman's retinas could detect.

Dr. J. Héricoult was yet more expansive in his own remarks on
the rays of Anna's clairvoyance.[73]

> And then came the discovery of X-rays, showing us
> that in effect 'absolute' opacity does not exist in it-
> self, and that in sum there is no material that cer-
> tain forms of energy, certain types of vibrations
> cannot penetrate; and then the question is posed, if
> the retina of hypnotized subjects does not under
> certain conditions become sensitive to unknown
> radiation.

It was known that Anna did not enter into a productive trance unless she was hypnotized by Ferroul. A subject-hypnotist bond existed between them that must be due to a common vibration which
awakened Anna's dormant responsiveness to the unknown rays.

According to the *New York Herald* and other American newspapers, Anna had "X-ray eyes."[74] The article described the marvel of
the first letter, attributing Anna's reading to X-rays while admitting
they couldn't pass through metal. At the end of the article a second,
yet more circumscribed test has been made, the results of which

---

[73] Héricoult (1897: 557)

[74] Here is a Woman with Eyes Like X-Rays, *San Francisco Call*, January 16,
1898: 23

had not yet been heard. Of course, they eventually were. J.G. Smith, writing in the *Proceedings of the English Society for Psychical Research* on some of the cases recently in the French *Annales*, in a stroke dismisses the *Herald*'s X-ray eyes label and the worth of more experiments with the occasionally clairvoyant Anna.[75]

A postcard printed during the Midi revolt of vineyard workers in 1907 shows the imprisoned Ferroul sitting on his cot in Montpellier prison while Prime Minister Clemenceau peers at him through the barred window. Among other inscriptions on the wall behind him is "A bas fraudeur," "Down with the fraud." That label did not stay with him very long.

Grasset did not mention Anna B. or Ferroul in his memoir of "occultism yesterday and today" published the following year.[76] He did use language both conciliatory and evasive on the subject of "human radiation." There's nothing to prove that human radiation equals psychic force, he wrote, and the multitude of radiations associated with human radiation (here he lists X-rays with others) perhaps prematurely, only tells us that there are more unknowns than knowns.

Moch's Xylope and Ferroul-Grasset's Anna B. were engagements of radiation relativity with human being in the relativity of the political moment. Both originated in France under the conditions emerging there, but made brief echoes in America where only the X-ray eyes and the deplorable X-ray phantasmagoria were evoked. This soon was repeated, with France making the broad suggestion and America seeing X-ray eyes.

---

[75] Smith (1898: 118)
[76] Grasset (1908)

6

FASCINATION

Where sight with X-rays was fictionally granted during the years immediately after their discovery the observers were men and the observed were women. The three variants of seeing with X-rays (eyedrops, spectacles and Xylopes) each centers around men who have acquired the sight and the women they seek to inspect. When the sight actually is attained, it is disappointing bones and "the gelatinous mass." The disappointment at the result, even singled out from Moch's Xylope sequence, is disappointment at not seeing the female "form divine" in the flesh beneath the clothing.

The female and male "form divine" had not long been visible in painted, printed, sculpted and cinematically unclad forms. The X-ray form was an unexpected ocular ambush of the flesh within the clothes.

"X-ray eyes", the phrase coming into common use in American English by early 1897, often was bestowed on women by male writers. It was metaphoric: a little Philadelphia girl's X-ray eye detected an eavesdropper through the thick curtains of her house's window.[77] A report of the discovery of a woman with X-ray eyes in a specific city was followed with a punning remark, for instance women in Washington, D.C. have an "eye for an X raise" [rise in pay] in "that enterprising town."[78] More often the woman with the X-ray eye can see through her husband, his plans, excuses and so

---

[77] One Girl's X-Ray Eye, *Morning Times* (Washington, D.C.) January 2, 1897: 6. Police confirmed that the eavesdropper had been there from footprints in the new-fallen snow.

[78] *The Times* (Richmond, Virginia) November 7, 1897: 4

on.[79] The nervous man approached by his wife wearing an X-ray lorngnette in McDougall's illustration is a variant, the X-ray mind-reader.

The grammatically conditional use of the X-ray eye phrase began in the early 1900s and continues to the present day. For instance, a thief *would have to have* X-ray eyes to find the location of a woman's pocket and the purse within it (given the flowing fabric of women's clothing at the time).[80] Men, women, children and people of many different occupations must have or would have X-ray eyes to accomplish a hypothetical end. This gender-free mode of X-ray eyes is entirely rhetorical, and independent of claiming or proposing or imagining what it actually would be like to see with X-rays.

Mrs. Walden, a migrant to New York from California in 1899, had few patients in the city that year. Only after she spent a period in Paris and there treated named aristocrats, including the Prince of Wales, was her name attached to X-ray eyes and success in healing. The opera singer Emma Thursby, returning to New York from Paris, passed on word of Mrs. Walden's coming to New York.[81]

> Mrs. Walden claims to see through the human frame and to thus ascertain the disease from which the patient is suffering. Then she proceeds to the treatment, which is simple, consisting merely in stroking and rubbing.

Mrs. Walden doesn't take the pulse or use other diagnostic techniques; she only looks at the patient. Emma Thursby is not recounting personal experience with Mrs. Walden. She "has been told" that she can raise relieving blisters on the patient's skin just by pressing down heavily with her hands. She does not believe that the Prince

---

[79] E.g., *Minneapolis Journal* April 16, 1906: 4. A French girl in Paris with X-ray eyes.

[80] Lee Jefferson, Finding Her Pocket, *The Tribune* (New York City) October 2, 1904: 18

[81] Her X-Ray Eyes Look You Through, *The Evening World* (New York City) October 26, 1900: 5

of Wales was one of her patients, but knows that Mrs. Walden has cured other eminent people of their troubles. The son of a French countess had not spoken for many years before Mrs. Walden took him under her care.

The woman is not striking in appearance nor is she well-educated, Thursby adds, but she has effected some "very remarkable" cures. In spite of her apparent lower-class origins, she has treated and healed aristocrats.

On her first round through New York in 1899 Mrs. Walden found that she was one in a population of non-medical diagnostician-healers seeking patients in the greater urban area. An 1874 publication gave the addresses of 9 "medical clairvoyants" in Boston and New York, 7 of them women.[82] The number and variety had grown considerably since then.

A medical clairvoyant was the 19th century category of healer who could detect by sight or touch, and remove, foreign substances from the body of a sufferer. Some American clairvoyants used the postal system, offering to return a diagnosis by letter if a postage stamp was included with the written query. Some, like Mrs. Walden, after locating the source of a client's malady, removed the offending internal substance by drawing it out in the form of a blister, an established medical procedure of the time.

Mrs. Walden seems to have acquired a framework to distinguish herself from her competitors during her sojourn in Paris. In imitation of European practitioners she made her clairvoyance into X-ray eyes and attracted upper classes hungry for novelty and health. Opera singers and ballet dancers, at least those who had achieved a level of renown, moved among the upper ranks of society: Emma Thursby was a credible emissary who with a slightly critical air could herald Mrs. Walden's triumphant return to New York. The newspaper listed several prominent New York society women who were treated by Mrs. Walden. After that, nothing more is heard of her.

An ecumenical clientele in Paris (H.R.H. The Prince of Wales! allegedly) devolved into an all-woman clientele in New York. Mrs.

---

[82] Babbitt (1874: 164)

Walden was the harbinger, or the sustainer, of a medical clairvoyance fixing X-ray eyes curatively rather than intrusively upon women. Evidence of a further development at least in the Manhattan form of this vision came six years later, as the result of a sting operation.

At the turn of the century medical societies formed by physicians undertook their own investigations to ensure that laws against practicing medicine without a license were enforced. There was an especially wide field for enforcement in the boroughs of New York City, where medical clairvoyants were only one type of practitioner charging the public for their diagnoses and treatments.

On April 13, 1906 three agents of the county Medical Society gave evidence before Magistrate Finn in Manhattan that Mrs. M.E.B. Frank and Madame Jane Endor were conducting a "hypnotic beauty parlor" and identifying illnesses without medical credentials.[83] Nearly two years later two women detectives again arrested the apparently incorrigible "girl with the X-ray eyes."[84]

Mrs. Frank billed herself as a Doctor of Suggestive Therapy. In her pamphlet, "A Word for Women" she assured women that the desire to be loved is as natural as breathing. "All the attributes that men find so alluring," of face, character, form, and movement can be acquired by the occult method Mrs. Frank teaches. Hypnotism, personal magnetism and suggestive therapy contribute to attaining this enchanted state. Suggestive therapy was a less insistent form of hypnotism directed by quiet speech.

A further level of training beckons to those who have gained the fundamental attributes. Mrs. Frank has the secret of turning them into human X-ray machines "which enables them to see through stone walls and look into the human body and tell at a glance if all the machinery is in working order." The machine language of Mrs.

---

[83] Girl of the X-Ray Eye, *New York Times* April 7, 1906: 9; Dark Days for X-Ray Eyes, *The Tribune* (New York) April 7, 1906: 10; Bad for the Girl with X-Ray Eyes, *The Salt Lake Tribune* April 14, 1906:8

[84] Women Who Hunt "Quack" Doctors, *The Daily Tribune* (New York), January 19, 1908:8.

Frank's prospectus was intended to appeal to the modernist impulses of her customers.

One of the medical society agents, Anna Molinelli, arranged for an introductory session with Mrs. Frank. She was seated in a room facing a "colored maid" over whom Mrs. Frank made a few hypnotic gestures. The woman said she was in "a beautiful place," told Molinelli what was wrong with her and collected the $2.00 fee. There is no further detail in the newspaper account.

Another of the agents went to the room of Madame Jane Endor, "the Girl with the X-Ray Eyes" at 1947 Broadway. The room's atmosphere, the agent tersely reported, was so thick with cigarette smoke that X-ray eyes would be needed to see through it. Her pupils dilated, the agent believed with belladonna, Madame Endor fixed her eyes upon the agent and uttered her insight.

> You have swallowed something alive. It looks like a
> fish. It's got legs like a spider. If you do as I tell you
> and come back for more medicine, you will be all
> right.

Madame Endor told her client that the many-legged creature was blocking the circulation of her blood. Magistrate Finn bound Madame Endor over for detention in advance of trial: she could not post the $200 bail. She was found guilty a few days later and fined $150 or sixty days in jail.[85] Mrs. Frank was paroled in the custody of her lawyer, thus preventing her, the unkind reporter continued, from putting her ability to see through stone walls to the test at the city prison, The Tombs, before her trial the following Friday.

Madame Endor, endowed with the name of the Biblical witch Saul consulted, seems to have been the recipient of Mrs. Frank's promised advance on charm and to have gained the X-ray eyes of her sobriquet. Her insight is routine medical clairvoyance with emphasis on the foreign object in the body projection that Mrs. Walden was using earlier. Many-legged invaders beneath the skin al-

---

[85] Illegal Practitioner Convicted, *New York State Journal of Medicine* 6 (1900): 297.

ways have sensorial appeal as an explanation for a confused and searching sufferer. The sight of Madame Endor's dilated pupils certifies the power of Mrs. Frank's regimen. Her fascinating eyes must have penetrating sight best associated with X-rays.

The name of belladonna connotes what it promises the "beautiful woman" user. It also is the species name of the plant, *Atropa belladonna*, from which the pupil dilating extract is made. Deadly nightshade and other names of the same plant more than hint at its uses. Perhaps its striking black berries against the backdrop of curling leaves suggested that nature intended it to make eyes appear wider.

Materia medica and therapy manuals described its many applications, always beginning with its effects upon the eyes: "inflammation, amaurosis and phantasms".[86] The notable dilation of the pupils lasted for several hours and was not affected by changing level of lighting. Dilated pupils had long signaled innocence and interest in the person gazed upon, with a shiver of awe gratifying to a male subject. It was an allure worth risking the inflammation and loss of sight that accompanied it. The phantasms, "sparks, colors and rays" would give the belladonna user the impression that she was seeing invisible light, making her diagnosis of the person watching her as she X-rayed them all the more convincing.

The cigarette smoke lowered the illumination and also caused the pupils of the client to widen. Madame Endor was an example of what Mrs. Frank's method could provide a woman, including a livelihood as a clairvoyant who looked the part.

Beauty advisers were unanimous in discouraging women from using belladonna to brighten their eyes, and from accepting the myth that actresses and society women used it regularly to gain their outstanding looks.[87]

---

[86] Hempel (1859: 1,356)
[87] Katherine Morton, The Folly of the Extremist; The Girl You Would Like to Be, *The Salt Lake Herald* May 6, 1906: 8.

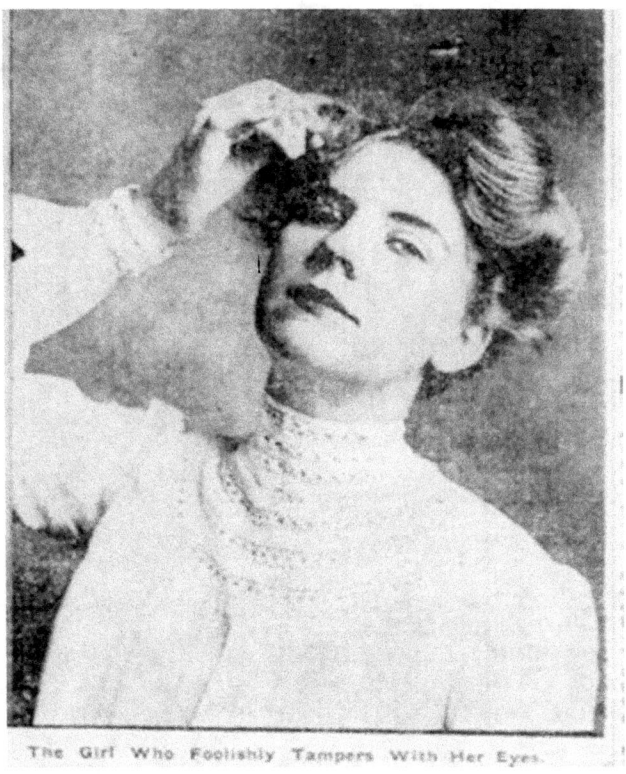

The Girl Who Foolishly Tampers With Her Eyes.

11. The Girl Who Foolishly Tampers With Her Eyes. Fn87

Bottles of eye preparations were printed with text denying that belladonna was an ingredient. Amid these warnings and disclaimers, belladonna had the appeal of danger, of wild beauty.

The combination of new beauty and medical clairvoyance Mrs. Frank franchised and Madame Endor practiced fit together well under the rubric of X-ray eyes. Anna Benzecry, the medical society agent who visited Madame Endor and gave testimony before the magistrate, had a long career of mimicking illness to receive treatment from unlicensed practitioners, both men and women, from herb doctors to Christian Scientists, but after the 1908 repeat encounter

12. The Shielding Shadow (1916) poster. Fn89

with Madame Endor she didn't close another X-ray eyes beauty parlor.[88]

Inducing X-ray eyes in paying customers did become part of the international repertory of beauty purveyors. The full, engulfing eyes inherited from the belladonna tradition-however they were accomplished-remained an accompaniment of the beautiful woman. X-rays weren't always evoked. In the publicity for a 15-part film serial, *The Shielding Shadow*, released on October 1, 1916, the staring dilated pupil eyes of a groping shadow that interferes protective-

---

[88] Virginia Sloane, Mrs. Francis Benzecry, Nemesis of Fake Physicians, *San Francisco Call*, January 19, 1913: 19.

ly in the life of the heroine are juxtaposed to her open-eyed face.[89] Two forms of X-ray eyes are illustrated together.

The film was the work of the same writer-director team who two years earlier had produced *The Perils of Pauline* serial featuring the ever-menaced Pearl White in the lead role, but it did not do for penetrating eyes what *The Perils* did for cars and trains. This poster may have been the first juxtaposition of the word "superman" with the phrase "see through walls."

The mix of beauty, fascination and all-seeing X-ray eyes in prose and picture did not pose the question of subjectivity. The ability to look into people and be seen as fascinating deferred the question of what the looker saw and felt about what she saw. What did the girl with the X-ray eyes actually see while she was being seen?

The Bulgarian writer Svetoslav Minkov had an answer. His short story *Dama s rentenovi uchi*, *The Lady with the X-Ray Eyes*, was published in 1934, and an English translation from the Bulgarian state publishing house appeared in 1965.[90] Minkov used the Bulgarian expression "rentenovi uchi,""roentgenic eyes" (which itself did appear occasionally in English) from "Röntgenaugen," the only word for "X-ray eyes" in German. A librarian and diplomat (the first Bulgarian ambassador to Japan), Minkov made translations of tales and other fiction into Bulgarian from German originals.

Mimi Trompeyeva has had the misfortune of being born with permanently crossed eyes. She seeks the aid of Maestro Cesario Galphone whose Cosmeticum Beauty Parlor for Ladies' Plastic Surgery serves women of the highest social circles desiring to improve ill-favored faces and bodies. Maestro Galphone promises to correct the trifling deviation from normal anatomy of her breasts with a paraffin injection. Her eyes, he says, will swim in radiant brilliance after he has treated them with a few drops of his new formula, Rentgenol.

Mimi's life after Rentgenol is set forth in her diary entries. She knows the acclaim that her eyes bring from people in the street. At

---

[89] The top half of the advertising poster from the *Evening Public Ledger* (Philadelphia) February 26, 1916: 5

[90] Minkov (1965: 113-20)

first she is frightened when she finds that she can see through solid matter to the skeletons and internal organs of people, looking into an animated graveyard. Then she discovers that by focusing on the heads of her acquaintances, especially the upper classes, she finds no brains at all there. Literally. She wonders how Jean, a rich and athletic gentleman who entertains her during a party, can be without a brain. Perhaps he and all the others think with some other part of their anatomy.

She discovers that the aristocrats do have brains of a sort that distinguish them from the common people. They are formed of a thin gossamer substance practically invisible. Mimi is overwhelmed by the attentions of Jean, and abandons herself to the rapture of her marriage.

Mimi at first has the same reaction as Herbert Newton and Moch's commentators to the sights seen through Roentgenic eyes. Instead of disgust and revulsion, curiosity and joie de vivre intervene and she throws herself into the feast of sight. Mimi can't separate herself from the brainless upper classes she detects, and she ends out investing herself fully (and maritally, given the size of the wedding settlement Jean's millionaire father bestows) in the life her roentgenic beauty has made possible. She is like the Xylopes who learn how to use trees they see through to the sap.

The subjective interior of the woman endowed with X-ray eyes is for Minkov as hollow as the hollow interiors Mimi is now able to see. Her vision doesn't grant her critical insight, only social advantage, and that because she loses her vision's interior in the gratifying exterior of her eyes' sight. In her last entry she and Jean are on the honeymoon heading toward Paris where she plans to buy an ermine coat and a ball gown.

Minkov has drawn Mimi after the frivolous female protagonists of novels. He asked himself: how would one of these women respond to being endowed with a great critical insight? Instead of sinking into depression and madness at the true sight of her social class, however, she embraces the rapturous marriage she has obtained through the improving applications of the beauty doctor. The only looker she can become is the one looked at.

This manner of X-ray vision continues in the rhetoric of the dating advice literature, occasionally with reference to X-rays he or she brings into focus on the body of the intended, or which the other party deploys to their satisfaction.

7

UNDERGROUND

A notice of the 1910 wedding of Guy Fenley of Uvalde, Texas to one of the Person sisters named him "a prosperous young man of one of Uvalde's old and prominent families." [91] The twenty-three year old groom brought the bride to his family ranch twenty miles away from town. The report did not say that ten years earlier Guy was widely known as the "the boy with the X-ray eyes" and that was the source of his prosperity.

The story was repeated in Texas newspapers and books on the lore of the Big Bend country that Justin Fenley was walking over a pasture on his property toward nightfall with his young son Guy when the boy said he saw water underground. [92] A reliable water supply was a necessity for anyone trying to build a ranching business on the dry flats far from the Rio Grande and its tributaries. Rather than going to the cost of sinking a well just on the word of a boy Justin Fenley tested him by placing a bucket of water beneath a wooden table and asking him to locate it. Guy's success in finding the bucket led his father to bring him back over the land to look for water. An aquifer did exist at the place and the 200 foot depth Guy indicated.

At Thomas Devine's ranch in the same area Guy found water at 175 feet down, said it was flowing in a southeasterly direction and

---

[91] Happily Married-Double Wedding in the Lone Star State, *Times and Democrat* December 3, 1910: 4

[92] The earliest account of Guy's abilities is in a letter from Austin, Texas circulated in western newspapers,e.g. He Had X-Ray Eyes-Remarkable but Well Attested Powers of a Fourteen Year-Old Lad in Texas, *Deseret Evening News* (Salt Lake City) April 20, 1901: 2, 12

pictured the intervening strata. His services came into demand among locals hoping to center their cattle-raising enterprises on wells Guy could locate. He reputedly did not accept money in advance of making a search, and was credited both with finds and with passing several bucket-under-table tests like the one his father first made.

Other fluids hidden within the earth, these not flowing, also fell within Guy Fenley's ken. Fenley's cousin D.O. Wells in Prescott, Arizona wrote to the boy's father to ask him to come to Arizona and gaze into the earth to find oil.[93] According to the *Brownsville Daily Herald*, in September, 1901 he actually did find oil in the Brownsville area.[94]

For a fourteen year-old boy to travel the distance from Uvalde to Brownsville, on the Gulf Coast of Texas, is a measure of the extent of his fame and the confidence of his hosts. From the first articles appearing in Texas and contiguous states newspapers in early 1901, to an informative notice in the *New York Times*[95] in December of that year the news of Guy Fenley, led off by the X-ray eyes phrase, spread through the entire country.

A wistful invitation to Guy asking him to come to their town printed in the newspaper of an Oregon mining community is among the final traces of his fame's progress from Texas outward.[96]

How did Guy Fenley become the "X-ray eyes boy" of brief, incandescent renown? The expanding circles of his fame are a study in reputation in the emergent natural resource economies of the western U.S. states.

At first Guy was a water diviner, one of several in his own area and part of an ancient tradition extending back to Europe and the Near East, all the more remarkable for seeming to come forth spontaneously in answer to his family's and community's needs.

---

[93] The X-Ray Boy Has Relative in Arizona, *Tombstone Epitaph* May 19, 1901: 1

[94] X-Ray Eyes Boy, *Brownsville Daily Herald* September 14, 1901: 1

[95] The Boy with X-Ray Eyes, *New York Times* December 14, 1901: 3

[96] X-Ray Eyes Boy, *Sumpter Miner* July 20, 1904: 4

He never was termed a zahori: the water-witching traditions of the Anglo population may have combined with the Hispanic-Arabic substrate, where eyes alone found the water. Zahoris certainly were known in Mexico and Central America. Guy saw underground water most clearly in the dark, unlike the zahoris, who saw in the light. The effort exhausted him, requiring sleep afterward. These attributes alone would place him in the category of precocious diviners.

In 1897, when he was 10 years old, his ability to pick up the glow of water in the earth became associated with X-rays for him but not for anyone else with comparable abilities.[97] Around this time he told his father he could see the skeletons of farm animals through their flesh. He did not claim to see skeletons of living humans.

Only one story gives Guy powers more in keeping with clothing penetrating X-ray eyes. He reportedly looked into the trousers of a man and saw a pocket watch he knew another man was missing.[98] The consequent disturbed relations between the two men probably discouraged Guy from further detective exercise of his eyes.

There is a hint of how Guy acquired X-ray eyes in a statement by Wigfall van Sickle, a Texas state legislator on whose property Guy Fenley found a water source: "That Guy Fenley, this fourteen year old boy, is possessed of an X -ray sight cannot be questioned." This was quoted in the 1901 letter, when Guy was 14. A knowledgeable and eminent person linked Guy's gifts to a new technology unknown beyond verbal descriptions in urban newspapers. The connection made, Guy himself saw animal skeletons, but avoided the indiscretion of seeing human skeletons, as he assumed the capacity of X-ray seer.

Besides letters from prospectors and promoters interested in bringing him to their lands, the local postmaster received requests from oculists asking to examine Fenley's eyes in search of the physical basis of his sight. At the height of Fenley's renown in 1901, John

---

[97] Other water witches in Big Bend country during the early twentieth century did not have an x-ray reputation, Miles (1976: 77-89).

[98] *Chicago Herald* cited by Mass Reformatory, *Our Paper* 21 (1906: 478).

Nance Garner, an Uvalde resident then serving in the Texas House of Representatives, received a letter from Richard Hodgson, LL.D., a researcher for the American Society for Psychical Research, asking Garner for "accurate accounts" of Fenley "from a scientific point of view." Garner replied as follows.[99]

AUSTIN, TEXAS, Feb. 8, 1901.
DR. KICHARD HODGSON.

DEAR SIR, In reference to the published newspaper account of the wonderful and unexplained gift of Guy Fenley, who can see water at any distance under the earth's surface, I have to say that my experiments have convinced me that he possesses this power of sight. I know that he can see through any substance and locate water beneath; also that he has located a number of good water supplies in localities in West Texas where water is almost an unknown luxury. There are many responsible people of Uvalde, Texas, and other places in that section of the State who have seen this wonderful sight displayed, and know that there is nothing mythical about it. I shall be glad to give you any detailed information concerning this boy and his wonderful gift, if you will inform me definitely as to what you desire on the subject.

Yours truly, JOHN N. GARNER
(Member of House of Representatives).

Hodgson monitored newspapers for instances of remarkable phenomena susceptible to scientific testing. He wrote to Garner, the Texas state representative from Uvalde, when news of Fenley's insights became known nation-wide. Garner was at the beginning of a long career in public life that would eventually lead him to the Vice Presidency (1933-41). In his response to Hodgson he vouched for Fenley's gift and affirmed his standing in the community without making reference to "X-ray eyes". He offered further information if

---

[99] Barrett (1913: 45)

Hodgson spelled out precisely what he wanted to know, which Hodgson never did.

The X-ray designation conveyed news of dowsing ability which if reported at all was a curiosity credulous locals subscribed to. Garner affirmed Fenley's gift for its utility to ranchers and farmers in West Texas, without repeating the newspapers' label. This was, in the spirit of all later comment on Fenley, a general study of water-witching in Europe and America put it, an instance of the "numerous instances of his X-ray gift, or whatever it may be."[100]

Fenley retired from active sighting of underground deposits in the mid-1900s, saying that he wanted to avoid being besieged by demands. He married and spent the rest of his life serving as county clerk in Zavala County, Texas. Apart from news of his wedding, there was no further mention of him in newspapers after 1905, but he did endure in compilations of the lore of the Big Bend country.[101]

Fenley was unique among water-oil-mineral diviners successful enough to be noted in the literature on the subject in the association of his special sight with X-rays. Fenley's own career and retirement parallel the rise and decline of water-witching in newly developing parts of America. There is no evidence that others who offered to find water, oil or minerals by looking into the ground claimed or were ascribed X-ray eyes.

An area similar to West Texas in having an energetic immigrant population eager to exploit the land but greatly needing water, the veldt of South Africa, also generated its dowsers. In the same article that discusses Fenley, W.F. Barrett of the Society for Psychical Research summarizes a letter from a man in the Orange Free State (translated from the Dutch) telling of a boy who can see the gleam of water beneath the dry earth.[102] At first the Afrikaans farmers feared this might be *backveld*, a demonic influence meant to lead them astray. As the authenticity of the boy's water-finding became established they designated it a gift. X-rays were not brought into play.

---

[100] Barrett and Besterman (1926:197)

[101] Smith (2011)

[102] Barrett (1913: 44)

This instance would not have been anything more than evidence for Barrett's weighing the psychical against the physical theory of dowsing if it didn't point to a history. Barrett did not group the Afrikaaner boy with the zahoris he had earlier considered. Anyone able to detect hidden resources underground possessed an extraordinary ability. For Barrett, as for the farmers, the question was the nature of that ability.

When Robert Deindorfer, a reporter for *Life* magazine, asked the Afrikaaner Pieter van Jaarsveld if he found water, oil, diamonds and gold because of his X-ray eyes, van Jaarsveld denied that X-rays were the reason. His retinas were very sensitive to vibrations from within the earth, the young pilot explained. He saw water as a shaft of moonlight, diamonds as flashing white spots, gold, oil and coal as characteristic shades of black. Yet Deindorfer titled his article for the magazine, "Pieter's X-Ray Eyes."[103] Pieter refused Deindorfer's label and defaulted to the time-honored dowser's sensitivity to mineral vibrations. His ability had to reside in "X-Ray Eyes" for the readers of *Life* magazine.

An article two years earlier in *Time* magazine had cited Pieter's moonlit water and shadow mineral lyrics of the underground, but no X-ray eyes.[104] Their attachment to Pieter's name in the *Life* article persisted in subsequent written accounts, mostly books on dowsing and supernormal abilities. Pieter van Jaarsveld seems to have been rejuvenated in several accounts: he is described as a twelve year-old boy as of 1963, which would place his birth date two years after the 1949 *Life* article in which he is a young man with an established dowsing business, Eureka Indication, Ltd. Primordial possessors of X-ray eyes often are children.

Qualifying someone who finds underground water or minerals by sight as having "X-ray eyes" is an imposition from the outside. During the early years of the spread of X-ray knowledge and apparatus there were a number of striking photographs making the formerly invisible insides of solid objects visible to anyone. There were no X-ray photographs of the interior of the earth, as useful as they

---

[103] Deindorfer (1949)

[104] Southern Rhodesia-Moonlight, *Time* December 1, 1947

would be to prospectors, engineers, and resource seekers. At the beginning a Guy Fenley could claim this scope of vision and gain fame, from which he retired with no ceremony.

It became clear to many who might employ them that X-ray photographs could not be made of the earth's underground. The psychomechanical sensing of the layers beneath-dowsing and divining-continued in mind if not in practice as ground penetrating radar and other geophysical instruments offered much more credible, if much more expensive, investigative methods. X-rays eyes remained a simple classification for anyone who could see into the earth. Pieter van Jaarsveld rejected it but it was imposed on him just the same.

After van Jaarsveld others in South Africa did not hesitate to claim the X-ray eyes label, and were brought to law when the water they were paid to find did not materialize.[105] The earlier boys with an X-ray eyes reputation existed in a milieu where that ability was a stream of hope. van Jaarsveld denied it because like the zahori expression in 16th century Spain, X-ray eyes in South Africa moved toward being a fraud and nothing more.

In the 1940s a Canadian businessman, J. Raoul Desrosiers, claimed to be able to locate underground water by an aching sensation in his ribs when he passed over the source. He also received the X-ray eyes label though he never said that he saw the water.[106] The label certainly is in the air and available for use, mostly in retrospect. The conditions of its use with Fenley are irreproducible.

When the folklorist Mody Boatright included a section on "the x-ray eyed" in his *Folklore of the Oil Industry* he included Fenley and van Jaarsveld among others who used the X-ray eyes label to explain their ability to find oil and water in the ground. Like the zahoris who in the Americas became magicians among others, they

---

[105] Lourens v. Genis 1962, ended in a judgment against the plaintiff, who according to the court exhibited "unreasonable stupidity" in believing an X-ray eyes water finder's claims. Bhana, et al. (2009: 347)

[106] Edwards (1966: 117-18). Edwards was the host of the radio program "Stranger than Science" during the 1940s and '50s. He does not give sources for his profiles of dowsers and prophets.

might just as easily have used divining rods or electrical devices to front their gifts.

The American mining engineer John Hays Hammond, arriving in South Africa in the early twentieth century to lend his expertise to gold exploitation, was proclaimed "the man who looks with an X-ray eye into the ground, as it were."[107] Neither in his *Autobiography* nor in other writings about his career did Hammond claim or receive this name. It was a casual designation imbued with the optimism of colonialist natural resource beneficiaries. As it were.

---

[107] *Copper Curb and Mining Outlook* 9(1911): 15

8

SEEING THROUGH WALLS

A novel by the French pulp writer Guy de Téramond (François-Edmond Gautier de Téramond) illustrates the evolution of fictional penetrant vision into the popular trope of X-ray eyes. The title of *L'homme qui voit à travers les murailles*, *The Man who Sees through Walls* (1914) gives away the plot sooner than the novel itself does: Lucien Delorme, a young man from the provinces staying in a Parisian pension sees projected on the wall of his room a pair of skeletons, each with a bullet in one part of the body. The pair set about strangling a third skeleton then remove objects from her person.

Delorme's telling this "dream" leads the chief of detectives to suspect that he was involved in the murder and robbery of a wealthy American woman staying in the room adjoining his. While he is not wearing his dark spectacles or looking through refracting glass Delorme sees through walls and into people. He spots the giveaway bullets in the skeletal shadows of the Comte d'Abazoli-Viscosa and his "Hindoo" servant Nam and identifies them as the murderers.

Late in the novel Delorme tells a Baron whose aid he is seeking how he came by his remarkable ability. Just after receiving a radiation treatment for a red spot on his nostril he stumbled in the street, fell and lost his spectacles. Suddenly he was surrounded by skeletons and could scarcely help himself. He was placed in a carriage and returned home. A doctor, who sees Delorme's eyes as two phosphorescent orbs in the dark after he removes a facial compress, elucidates how Delorme's fall after the treatment precipitated his altered vision.[108]

---

[108] Téramond (1915: 300)

" 'Wait. This atom of radium was drawn in, carried along by the circulatory stream. The violent shock of your fall localized it in your brain, at the end of some vessel without an outlet. Your skull has become a radiographic apparatus. You see with X-rays!'

Realizing that his X-ray sight brings with it an inevitable degeneration of his entire body as the radium multiplies and spreads, Delorme carries back to the villainous count a bomb which is then detonated by the servant, obliterating all three.

Lucien is the innocent victim of the Count's criminal machinations. He is seduced by the Count's beautiful accomplice Juliette (calling herself "Georgette"), whom he sees at her work almost too late.

Guy de Téramond was the author of erotic and adventure novels, travelogues, novelizations of films and the occasional libretto of an opera. In 1906 he edited and wrote the text for a collection of photographs, of art works and living people, on the beauty of the nude in antiquity, religion and modern life. Strange happenings in exotic locales mingled with romance and/or sex form much of his writing, which has disappeared entirely or never entered library collections.

He kept his novels of the thrills of the flesh separate from the more modest thrills in his adventure novels. *L'homme qui voit à travers les murailles*, which one English reviewer, with an eye to the other novels, called "as innocent as an alphabet card" went through at least six French editions between 1914 and 1926. It was translated into English by Mary G. Safford and published as *The Mystery of Lucien Delorme* in 1915, the only Téramond novel to appear in English. In 1919, Safford's translation was serialized in the *Washington Times* newspaper under the title, *The Man with the X-Ray Eyes*. In this final step of the novel's journey to America no plot content was changed, but the illustrations were.

The four illustrations by J. Henry in the book-form Safford translation are dramatic scenes: a body is found, the Count is bewildered, Lucien is in intimate converse with Georgette (2).

The Man with the X-Ray Eyes newspaper serialization ran for 30 installments between January 30 and March 19, 1919, always appearing on the first page of the magazine section. Three of the four illustrations were by Julia McCarthy and none of them was of a scene chosen for the illustrations in the book. The February 19 installment concerns Lucien's escape from a safe in which the Count and Nam have locked him after he has confronted them with the murder of the American woman. Safecrackers executing a carefully planned burglary inadvertently release him.

The steel plate yielded, the safe opened, and Delorme sprang, to his feet, confronting the burglars.

13. The steel plate yielded, the safe opened, and Delorme sprang to his feet, confronting the burglars. *The Washington Times* February 19,1919: Section Two

Two parallel beams radiate from the eyes of the imprisoned Lucien as the burglars clear aside the wallboard. One of them falls backward and raises his elbow in astonishment as his comrade works at the opening.

In the novel the scene is related second hand, when the count learns from the police that the safe has been opened and the jewels that were supposedly kept there are gone. He had expected Lucien Delorme to die of suffocation and then be taken by Nam and thrown into the Seine, a bogus suicide note left to explain his disappearance. From Lucien Delorme's point of view he is imprisoned in the empty safe thinking his life is lost when the burglars break in and quickly scatter when they find a man inside. Lucien then makes his escape and goes into hiding. He does not see through the wall of the safe and the burglars do not see his glowing eyes.

The second illustration (March 19) places Lucien on the other side of the wall of a private room at the Silver Pike restaurant as Georgette inside the room removes a vial from her dress pocket and pours poison into the glass of champagne she will offer Lucien when he returns to the table. He has removed the glasses that block his see-through vision, and realizes that Georgette is not the innocent young girl he thought she was. Fleeing the restaurant still glasses-off, he discovers Nam in the disguise of an old flower seller.

These two illustrations narrate the scene from the viewpoint of the reader looking into the story. They translate the fictive fact of Delorme's X-ray emitting eyes into a picture of human radiation in action, not just a vision of what Delorme sees.

The novel's own account of what Delorme sees is a phantasmagoric skeleton show recalling the early films and the X-ray projector opening up bodies before its rays. This X-ray vision simply saw X-ray photographs of the object expanded into a world of X-rayed objects. The illustrations add a later state of X-ray knowledge: they are a directed beam and not general lighting. Now the eyes are a beam device corresponding to the change in title to "the man with the X-ray eyes." The doctor's assessment that Delorme's skull has become a radiographic apparatus is given explicit visual form with the X-rays pictured as light beams cutting through darkness.

14. He saw Georgette draw from her waist a vial and pour the contents into his glass. *The Washington Times*, March 1, 1919: Section Two

The graphic precedent for these beaming eyes is the Eye of Providence, a Masonic symbol showing a single eye surrounded by radiant lines. It was adopted as part of the obverse of the Great Seal of the United States in 1782, and became part of the iconography of the one dollar bill when the country began to issue national currency. This is the all-seeing eye of religious iconography, depicted as a single large eye or an eye projecting linear rays.

The rays imprint the power of universal sight that accompanies the light issued by a luminous body. Single rays or light paths emerging from eyes were used graphically to connote attention to a single element of a scene. A 1906 newspaper illustration of Leo

99

Brett, reputed at the time to have X-ray eyes, shows him issuing cones from his eyes toward a hand he is examining (see Chapter 9). The graphic convention probably was derived from line illustrations of X-ray tubes. The illustrator adapted this iconography to X-rays that pass through intervening layers to expose critical plot elements.

For the novel it is sufficient that Delorme's eyes merely glow in the dark when his spectacles are removed, phosphorescing as the radium in his brain does. The glowing eyes of "radioactive" beings was a trope of science fiction being set into place. Radium does emit X-rays in the course of decay, not with the intensity the fiction requires, and it was injected into and applied to patients as part of treatment for various conditions.

Delorme had his radium treatment to remove a slight blemish from his nostril: the dire results of a minor procedure reflect unease about the vanity use of radium in the body. Téramond was the only fiction writer to use radium to give someone X-ray eyes. Other writers used it to confer other powers. Delorme's sad fate in the novel, to be consumed by the proliferating radium as if it were a cancer, reflects the actual dangers of cosmetic radiation treatments not well recognized until decades later. The dangers of frivolous X-ray exposure were represented in fiction during the first year of Roentgen's discovery.[109] That did not at first lead to cautions about their use.

In the newspaper serialization, Delorme is not merely the mysterious man who can see through walls whose origins will be revealed in the last chapter. He is the man with X-ray eyes aimed at the burglars through a wall his vision might have torn open. He is the man with X-ray eyes who stops just short of his treacherous fiancée's body to capture her hand in the act of poisoning him. The eyeglasses that in this case inhibit his X-ray vision encircle his eyes in the first illustration, perhaps a reflexive if contradictory tribute to X-ray spectacles.

---

[109] Griffith (1896) tells the tale of a wealthy man who has his brain repeatedly X-rayed in an unsuccessful effort to photograph his changing emotions. He becomes a leathery-skinned, hairless outcast as a result.

Science writers and journalists might dismiss the technical possibility of X-ray vision. The enthusiastic drive that began the twentieth century with the infinite possibilities emerging from electricity, X-rays and radioactive materials, powered by known inventors like Thomas Edison and the assurance that what could be imagined was certain to become reality was not to be stifled by the principles of science. Coherent images with a strong psychosocial base trumped reason itself, and the man with the X-ray eyes acquired a body.

In making the leap to America, Delorme, like Anna B. and other immigrants, has acquired "X-ray eyes," both in name and in illustration. Though the concept certainly can be made clear in any language, the expression encapsulates an image, an American cultural glyph, of a person with a peculiar ability to see through objects and people in the same way a radiograph forms an image. Prior to this expression, seeing with X-rays was a result of the rays being spread over a scene, as in the early films. The editors who retitled Téramond's novel and the illustrator who gave the title visual form individualized the X-ray vision formula. They used a European mystery-adventure character to inscribe a glyph that had been culturally present for over a decade.

No film ever was made of Lucien Delorme's adventures. Other media standards were developing for the exhibit of X-ray eyes, linking them to hypnotism and spiritualism on the way to an enduring scientific-theatrical presentation supporting individual careers.

9

OTHER WORLDS

Invisible worlds were being progressively disclosed during the nineteenth century. Photographic emulsions with and without lenses made spirits and radiation visible. Attached to microscopes and telescopes, cameras brought tiny and vast reaches of space before the eye. Cameras froze motion to disclose the intermediate positions of a moving object and through animation projectors made fixed objects move. Spectroscopes made visible the sequence of colors within earthbound substances and astral bodies. X-rays casting shadows of interiors formed one more progression of visibility.

The new visibilities had their corresponding eyes, some invented before X-rays were discovered. Imaginative writers and scientists projected the possibility that there were yet more worlds to become visible and yet more eyes to see them. They began traditions that assimilated X-rays and other rays into general illumination. They postulated other worlds parallel to and incorporating the one discovered by seeing with X-rays.

"Un autre monde," "Another world" a story published by J.H. Rosny in 1895, points to a paranormal life centered on vision and motion.[110] Karel Ondereet is a Dutch boy born with a light purple skin and a sickly constitution who is enlivened when a maid feeds him alcoholic beverages. His eyes are covered with an opaque horny layer that does not obstruct his vision. For him the colors of the spectrum are muted, and a range of colors beyond violet with gradations between them are visible to him alone. He can see through paper and wood. Glass is black and water not transparent. He can

---

[110] Rosny (1898). English translation by Damon Knight in Hartwell (1997: 539-57).

look directly at the sun through clouds and he sees through walls. He moves and speaks so quickly that he is considered impaired.

He can reconcile his extended perceptions with those of everyone else by understanding that the names of colors are arbitrary conventions. He subsists in a relativistic universe, barely able to function according to the strictures of society, family, church and school. Orders of flat creatures of colors imperceptible to anyone else go about their lives before his eyes. He names the terrestrial creatures Moedingen and the aerial creatures Vuren. The infinitely flat Moedingen conform to the shapes of objects on the ground and the Vuren pass through everything as they float in the air. They are a harmless and entrancing spectacle that further alienates the wonderstruck Karel from others. He flees to the city, where he has the good fortune of being taken in by a doctor who offers him shelter in exchange for the opportunity to study him.

Karel's vision and speed of perception allow him to perceive physiological processes in human bodies not previously known, leading the doctor to develop diagnoses and treatments not possible by routine observations. Karel undertakes a study of the other world he sees all around him. His attentions give life to a fading young woman he encounters in the hospital. They marry, and their son has Karel's skin color and abilities. To Karel's great joy, succeeding generations with his unique abilities will perceive and examine the other world around them.

Rosny was two brothers, French-speaking Belgians of the surname Boex, who wrote a number of stories together before separating into Rosny l'aîné and Rosny le jeune in 1907.[111] The elder brother continued to write stories and novels combining detective, supernatural and science romance genres until his death in 1940. Rivaling his older contemporary Jules Verne in daring, Rosny conceived of an earth invaded by inorganic aliens that have no way of communicating with the primitive humans there, of a dying humanity supplanted by ferromagnetic creatures and of apparently alien activity that erases a portion of the spectrum, the part that carries heat to the surface of the earth. Traveling to Mars in a durable ship,

---

[111] Vernier (1975) is a critical survey of Rosny's science fiction.

humans encounter a threatened population of airy beings. One of the travelers precipitates parthenogenesis by a Martian female, the first recorded interplanetary act of reproduction.

Rosny was as occupied with human (d)evolution as he was with the science of light and seeing. Many of his stories find humanity headed toward extinction replaced by entities of knowable or unknowable intelligence and purpose. "Un autre monde," coming toward the beginning of his oeuvre, before the dashed hopes and wars of the twentieth century brought a pessimistic shift, looks forward to a new type of human who can live with the full spectrum of life on the planet. Rosny's stories take place in this world under the influence of hitherto invisible surrounding worlds.

Karel has ultraviolet vision and a sped up timing that allows him to see invisibly accelerated life forms in their own colors. He unites several of the accomplishments of 19th century science in his own being: breaking down light into constituent elements that reveal the composition of matter; time-lapse photography; the world visible only through the lens of a high-powered, illuminated stage microscope. What would later be called X-ray vision falls within the realm of his senses. Karel has evolved to incorporate the instruments of discovery.

Moch may have had Karel in mind when two years later he thought up the wooden-eyed Xylopes, who also were a positive further step in human evolution. Rosny anticipated the inability of those endowed with X-ray vision to see through glass or certain metals. The new variety of humanity fathered by the alcohol-powered Karel is the humanity of the twentieth century looking over their world with widening eyes.

The possibility that technology might be fashioned to achieve telepathy and clairvoyance for all stimulated many writers (and inventors). Confirmation that the earth is surrounded by a magnetic field and the sending of coded radio messages over terrestrial and oceanic distances contributed to a haste in the imagination to fully inhabit the larger world in which this is possible.

A story by H.G. Wells, first published in a periodical in 1895 and later included in the numerous anthologies of Wells stories also seems on the brink of a vision that can include X-ray penetration.[112] "The Remarkable Case of Davidson's Eyes" has fallen to the lot of a technician in a college research laboratory to narrate. Davidson is in another room, standing between the poles of a large electromagnet when a lightning strike the narrator hears but doesn't see leaves Davidson standing dazed and making odd remarks.

Over the following time it becomes apparent that he doesn't see his surroundings-he can collide with furniture if he tries to walk-but he does see a sandy beach where a beleaguered schooner is foundering directly offshore. He can hear the voices of the people presently near him. When he moves into the landscape, all inaudible to him, he passes through the dunes without feeling their mass. The beach scene dissolves in patches until his sight in restored.

Three years later, the narrator is a guest at a dinner at Davidson's house where a Navy lieutenant also is a guest. Davidson recognizes the ship in a photograph the lieutenant happens to show. The lieutenant confirms all the details Davidson witnessed. They took place during a naval expedition to gather penguin eggs from an island in the far south. A professor theorizes that the lightning strike temporarily placed Davidson's retinas into a fourth-dimensional wrinkle. He saw events taking place concurrently on the other side of the planet.

Davidson has experienced electromagnetically induced clairvoyance that dissipates as the charge diminishes. This might seem to be Wells sparking up a routine remote vision story of the kind that has been told since antiquity but now in the context of electromagnetism and the fourth dimension. A television dramatization of the story broadcast in 2005 makes Davidson a participant in a Royal Navy experiment in radio broadcasting with Faraday coils boosted by the accidental discharge of the lightning.[113] The Royal Navy experiments were introduced to add corroborative detail and ground the otherwise awkward or possibly supernatural intervention of the lieutenant.

---

[112] Wells (1904: 168-91)
[113] Williams (2007: 176-77)

Wells was, as usual, on a track that paralleled trends in invention and in cinema. He did think of Davidson's eyes receiving light rays from the remote south of the planet. The lightning induced such a powerful increase in the magnetic field that the light reaching Davidson's retinas was transposed from that remote place rather than from the immediate vicinity. The eyes of someone viewing that scene as it was taking place could serve as a trans-spatial movie camera that sent out electromagnetic rays to be received by a screen like Davidson's eyes.

X-rays were not named when Wells wrote his story. Yet X-rays were a conceivable medium for the transmission of images over great distances because they penetrated everything projecting an image before them. The other world revealed by the X-rays was one in which all places were potentially present in the same place.

According to newspaper reports, Boris Rosing (spellings varied) of the St. Petersburg Institute of Technology in 1910 invented a "roentgenic eye" using an "electrotelescopic apparatus".[114]

> With it, says the professor, an employee sitting in his office can see all parts of the building, or a theater performance can be watched at home; while generals may be enabled to watch the movements of an enemy as well as those of his own forces. The details for the moment are withheld.

Rosing had taken out Russian, English and German patents in 1907 for a construction of two polyhedral mirror drums rotating at different speeds that reflected a simple geometric figure onto a photocell which sent signals to a pair of modulating plates deflecting a beam to project the figure onto a photosensitive screen.[115] He was the first to incorporate a cathode ray tube as a receiver of phototelegraphic images. In 1911 he patented an improved version, and with the assistance of his student Vladimir Zworykin exhibited a

---

[114] Invents All-Seeing Eye, *Washington Herald* December 11, 1910: 1. Named "Prof. Rossig" in this article.

[115] Shiers and Shiers (1998: 49n349)

working model of the system, at the St. Petersburg Institute on May 9.[116]

The newspapers copied the rhetoric of penetrating and telescopic X-ray enabled vision from articles that Rosing wrote for popular French and German magazines detailing the workplace, military and entertainment potential of his invention.[117] Articles in general technical journals like *Scientific American* took up the language and the promise of "the electric eye" as Rosing framed it.[118] The "roentgenic eye" of the cathode ray tube and "the telegraphic eye" of the interrupted signal coexisted in accounts of the technology. The eye was the photocell transforming patterns of light into electricity that could be turned back into the original light pattern.

Rosing's design belongs to the history of " television," a Greek-Latin compound devised at the turn of the century but not consistently used as a label for technologies that transmitted visual images. There is a tendency to sight backward from a finished technology and write its history in terms of tendencies toward that working finality. Rosing was in a tradition that encompassed telegraphy, radio and ultimately fax technology, but he phrased the hopes for his device in terms of monitoring, surveillance and delivery of entertainment, of another world that would result from the electric eye's wide presence.

The St. Petersburg Institute of Technology, where Rosing continued to work on his apparatus in the years following the 1911 patent, was unusual among Russian state schools in not requiring a gymnasium education of entrants. As an institution it avoided the elitism that characterized other institutions of higher education, and it survived the 1917 revolution, changing its name to the Leningrad Institute of Technology (1924) and then back to the St. Petersburg designation after the city's name was restored (1991).

This focus on technical ability over dedication to the classics fostered a freedom to experiment in contact with developing Euro-

---

[116] Glinsky (2000: 38)

[117] Burns (1998: 121) translates a portion of an article in the French magazine *Excelsior* of which the newspaper piece above is in effect a summary.

[118] Grimshaw (1911). Shiers and Shiers (1998:54n400-12) list others.

pean trends. Rosing envisioned his electric eye looking out from a central place in an industrialized capitalist state. His futurist television was an X-ray eye extending the surveillance of the employer and the general over their subordinates.

This vision was sufficiently compatible with the aims of the Bolsheviks for Rosing to remain in place and at work as the Soviet system took hold. He published a book-length compilation of his studies in 1923[119] the title of which indicates his continued dedication to the ideal of an electric telescope. He eventually fell afoul of Stalin's regime and, accused of not conducting research in the service of the revolution, was exiled to Siberia in 1931 without a right to work. The influence of his supporters obtained him a place teaching electronics at an agricultural college in Archangelsk. He died not long after the transfer, in 1933. In the year before his exile he published in a German technical periodical a brief article on electrical fluctuations in photocells used in reading machines for the blind.[120]

Rosing's student Vladimir Zworykin after he completed his degree studied X-rays with Paul Langevin in Paris. In 1898 Langevin had discovered secondary emissions from metals irradiated by X-rays. He continued to explore this over the next decade as he rose in the technical school hierarchy and became the head of his own laboratory. Zworykin did wartime service in Russia as a supply officer, and in 1919 made his way to the United States. At RCA laboratories he designed the first all-electronic television camera, the iconoscope, which he described as an "electric eye" then a cathode ray receiver, the kinescope, in effect completing the equipment needed for an electronic television system.

Zworykin was one important contributor to the most widely used system of television broadcast and reception. The broadcast from the camera passed over distances and through walls but it was not carried by X-rays nor did it bear an image of what it passed through. The general principle of Rosing's electric eye was realized but not its closed circuit surveillance of buildings or its long dis-

---

[119] English translation of the title: *The Electric Telescope (Sight at a Distance)*. The book itself did not appear in translation.

[120] Rosing (1930)

tance views from a central console. Both aspects of Rosing's vision had to await the development of compact video cameras and satellite technology. Rosing thought in terms of electric current generated by photocells carrying light through walls.

A world made transparent by X-rays was at the root of early television. Members of the early Russian avant-garde, painters, sculptors and filmmakers, held theories of visible space and proposed projects dissolving exterior surfaces, walls and skin. Early television at least in the Rosing-Zworykin tradition of experiment was one of these projects. Imaginary television systems like the one enabling the comedian George Burns to spy on others by turning on his television set, or the wall-filling composite screens allowing fictional security organizations to scan city streets, desert battlegrounds and mountain slopes from space. These fictional systems are now being realized in urban security systems that link widespread cameras to central image consoles in command centers.[121]

We know about Rosing's *Excelsior* article because a quotation in an early German history of phototelegraphy prompted Zworykin to quote it in a book he wrote on the development of television.[122]

> The range of application of the telephone does not extend beyond human conversation. Electrical telescopy will permit man not only to commune with other human beings but also with nature itself. With the 'electric eye' we will be able to penetrate where no human being penetrated before. We shall see what no human being has seen. The 'electric eye' fitted with a powerful lamp and submerged in the depths of the sea, will permit us to reach the secrets of the submarine domain. If we recall that water covers three-quarters of the earth's surface, we readily realise the infinite extent of man's future conquests in this portion of his

---

[121] Singer (2012) on IBM's Smart Cities system in Rio de Janeiro

[122] Zworykin (1958) quoting in English translation from Korn and Glatzer (1906)

domain, till now inaccessible to him. From now on and in all future time we can imagine thousands of electric eyes traveling over the floor of the sea seeking out scientific and material treasures; others will carry out their explorations below the earth's surface, in the depths of craters, in mountain crevices and in mine shafts. The electric eye will be man's friend, his watchful companion, which will suffer from neither heat nor cold, which will have its place on lighthouses and at guard posts, which will beam high above the rigging of ships, close to the sky. The electric eye, a help to man in peace, will accompany the soldier and facilitate communication between all members of human society.

The exhilaration in Rosing's words reflects the inspiration leading Zworykin to carry the phrase "electric eye" forward into his own invention the iconoscope.[123] There is something of Rosny's Karel Ondereet in Rosing's electric eye being passed down to the next generation with his figurative son Zworykin. The rhetoric of the passage anticipates the hope of universal communication that some of the fashioners of the Internet also hoped for. This all-seeing eye has a tone of conquest and domination by oversight that brings to mind eye in the sky universal surveillance and the fears it excites.

The electronic dream of distance vision lingered in the fiction of X-ray eyes though it was not realized in the mundane wonder of television itself. Another dream world framed by X-rays was the very small, a vision of thought as microscopic movement.

Dr. Max Baff told a newspaper interviewer that he was in touch with a scientist in Buenos Aires who could fit an X-ray tube with a magnifying apparatus and film camera.[124] The resulting "quintamillamicroscopia" records brain tissue magnified 5,000 times. Changes in the pattern of blood flow correspond to thoughts initiat-

---

[123] Zworykin (1934)

[124] X-ray Moving Picture Machine Shows Brain at Work, *New York Times*, September 4, 1910: Magazine Section, 4

ed by speaking to the person whose brain is being examined. Baff affirmed that the brain of a genius would respond more quickly than the ponderous brain of an idiot. The power of the brain lies in the now observable process of blood replenishment in the neurons. The study of psychology, normal and abnormal, and the education of youth will be improved immeasurably by this instrument for detecting the temper of thought.

Baff, a research fellow at Clark College, had previously gained notice for his claims that Theodore Roosevelt exercised a hypnotic influence through his speeches, and that women were savage in wearing their hair long and adorning themselves with gold. He would later make national news in ironic small fillers, with his claim that he could scientifically select the ideal wife. Except for the women's savagery claim, which attracted entirely negative commentary from both men and women, Baff was prescient in his description of the magnetic power of politicians, attempts to technologize mate searches, and in his anticipation of venography and functional MRI's. Yet his understanding of cerebral blood flow was not even up to date for 1910. He did, however, conceive of X-rays used to magnify thought as a material process.

Several emergent technologies were initially conceptualized as forms of X-ray vision looking into other worlds at hand, the telescopically remote and the microscopically small from the confined locale of an ocular or screen. The ability of X-rays to render a whole interior summarily present before the eyes led to thinking they might be fitted to human eyes for the capture of other worlds.

The horny-eyed Karel anticipates the wooden-eyed Xylopes in the thought that these other worlds would be revealed if obstructive visible light were cut off from healthy eyes then made sensitive to ambient penetrating radiation. The use of a cathode ray tube, adapted from X-ray production, was a decisive step toward television technology. The cartoon fiction of X-ray eyes reading thoughts by gazing into the brain was brought closer to practice with the idea of X-ray microscope mapping of fluctuations of blood supply in brain tissue. Like their exposure of bodies that always were there unseen, X-rays become vision brought other worlds into view of this world.

X-ray eyes were a reflex that could not be stopped in this drive toward ever wider vision.

10

LEO BRETT, BEULAH MILLER, QUACKENBOS AND MÜNSTERBERG

Guy Fenley had his water finding made into the result of X-ray eyes where urban news intersected with West Texas folk traditions and Hispanic zahoris activated by drought. Leo Brett came into the gift attributed to X-rays around the same time but at a younger age, and given his environment, with a medical purpose. The American X-ray eyes trope, affirmation straining against denial, is exaggerated by the interest of doctors and psychologists.

Frank M. Brett, a Boston physician who taught bacteriology and "physical diagnosis" at the Boston College of Physicians and Surgeons, made it known in medical circles that his infant son had acquired a unique ability. In 1899 Benjamin O. Flower, an editor of the "magazine of constructive thought," *The Coming Age,* interviewed Brett concerning Brett's then eight year-old son Leo (Afley Leonel) Brett, who had been endowed since early childhood with "supernormal sight."[125]

> His father believes the sight to be in most respects analogous to the X-ray, with the important difference that the boy is able to recognize and indicate colors and shades of color, which, of course, the X-ray fails to disclose.

The gift first came to Frank Brett's notice after Leo, recovering from a bout of scarlet fever, was not absorbing nutrients from his food. The boy began to regain his strength after Frank Brett hypno-

---

[125] Brett (1899: 449)

tized him. While in trance Leo said he could see through skin to the bones. As an infant Leo had detected the presence of a woman's skeleton kept in a locked closet off his room. He repeated his baby talk word "oolin," which Brett took to mean "woman." This would be good performance for an adult anatomy student looking directly at the skeleton.

In trance, Leo looked through the clothing of a young man brought for diagnosis and saw the traces of the bullets that had shattered the man's arm. Leo saw the bullets as "holes," and impressed his father with his persistence in this reading despite being subject to hypnotic suggestion to the contrary. The man's arm had been repeatedly X-rayed, and he was able to confirm the count that Leo made of the holes in locations where bullets had been detected. Flower witnessed this test, which consolidated the idea that Leo's supernormal vision was analogous to, but not the same as, X-ray photography. The boy's insight developed in a medical instructional environment as X-ray images became a way of visualizing the human interior.

In the interview that Flower conducted with Brett after the boy had ridden off on his bicycle to join playmates, Brett told him that Leo saw both normally and supernormally, with his eyes open, and not closed like others exercising special sight. Leo only saw interiors when his father hypnotized him, not at other times and not when others hypnotized him, because Frank Brett implanted the post-hypnotic suggestion that Leo not "see" for anyone else. While in a trance Leo said he felt a "thrill like electric current" running through his body. He could not see on stormy days or in a crowd. Wood, thick paper, glass and most metals were impervious to his vision.

Leo performed a number of visual diagnoses which, his father emphasized, depended upon his sensing colors and textures which would not be realized in an X-ray photograph. He saw that a young girl had not swallowed a one-cent coin as her parents believed, but had a mass at the base of her stomach. He saw tumors, the location and trend of bone fractures, tubercles in the lungs and fatty masses in the intestines. Frank Brett, with his knowledge of anatomy and pathology, gave names and diagnostic meaning to Leo's observations. The only explanation of the vision that Brett put forward related it to enhanced sensitivity to vibrations while hypnotized. Otherwise, Leo was a normal boy.

In the aftermath of Flower's interview Leo's ability gained currency in newspapers and periodicals as "A Human X-Ray" or "X-ray Eyes."[126] Two newspaper reports on Leo printed six months apart during 1899 set out the semiotics of applying X-rays to his sight. The first article, published soon after Frank Brett's interview was printed, resembles it in deemphasizing the role of X-rays, as both Leo and Frank Brett did in the interview.[127]

MASTER "AFLEY" L. BRETT.

15. Master "Afley" L. Brett. Fn127

---

[126] A Human X-Ray, *Aberdeen Weekly Journal* July 17, 1899: 6; X-Ray Eyes Stephenson,ed (1904)
[127] Sees Thro' Clothes, *St. Paul Globe* July 2, 1899:24.

Leo did have a vision of the bones inside the body while under hypnosis, Frank Brett specified, first in July, 1897, when Leo was nine years old. The correspondent found "all the wonderful faculties of the Roentgen rays" in his sight, and in addition he can distinguish colors and textures critical to diagnoses. To the case of the girl who did not swallow the cent and others the correspondent added Leo's correct delineation of fractures in his own arm he had sustained as a schoolboy. In this account Leo not only sees bones and organs, but also the sparks in the brain that correspond to the nerves firing at the muscles of a man voluntarily moving his arm. A lithographic portrait of Leo centered around his staring eyes accompanied the article.

The second newspaper account of Leo, later and shorter than the first, is headed by the X-rays.[128] His ability generally to see bone fractures through clothes, and the case of the unswallowed cent are rehearsed, but in this account he father is not a doctor but a parent who consults medical men after the boy sees the inside of the father's hand while playing. There is no mention of Leo's infant identification of the woman's skeleton, nor of the anatomy chart-like reading of the bones and organs. The author of the article, who does not state he visited Leo, relates Leo's vision not only to X-rays but to the X-ray apparatus itself.

> The lad uses this power by so concentrating the sight as to shut out ordinary daylight. The air, he says, is then filled with flashes of a pale greenish light, which illuminates the objects to be examined. This light, he says, is the same as the X-ray in the Crookes tube. Daylight is then darkness or a reddish black.

In keeping with this description, there is an engraving of a formally dressed Leo training the beams of his eyes on a sleeved hand he holds.

---

[128] Has X-Ray Sight, *Hopkinsville Kentuckian* November 24, 1899: 7.

HIS POWERFUL VISION.
(Massachusetts Boy Whose Eyes Are Veritable X-Rays.)

16. His Powerful Vision. Fn 128

The two pictures of X-ray eyes are registered in these two articles: the staring face and dilated pupils of the trance state and the directed beams of the eyes as X-ray tubes.

Both Leo and the reporter seem to be familiar with the operational appearance of a Crookes tube, which is directly related to the boy's powerful vision made manifest in the beams of the illustration. Neither Frank nor Leo Brett mentioned the greenish light in the first-hand interviews previously cited, but it does show up in other accounts.

A footnote in a book on hypnotism by the New York physician R. Osgood Mason alludes to the "peculiar perceptive power" of the

South Braintree boy.[129] He does not mention X-rays except to re-mark that Leo Brett scorns them, and says that he sees clearly on his own. Mason did not visit the Bretts. He quotes Leo from an un-named source that while in trance and performing a diagnosis he saw "an atmosphere of pale-green light flashing in every direction about the patient to a distance of four or five feet."

Mason attributes Leo Brett's light to the "odic force" shared by electricity, magnetism, heat, light and chemical reactions according to the German occult philosopher Karl Reichenbach. The greenish light that enables Leo to see into bodies is for Mason a visible mani-festation of the vibrations shared by all energy forms and by living beings under hypnotic influence. Leo's eyes are attuned to this light as were the eyes of Reichenbach's sensitives.

The glow of an X-ray tube while charged, and of the platinocya-nate screen often the target of the invisible rays, was pale green. Numerous sources state this during the first decade of X-ray genera-tion. It was so widely acknowledged that for instance there was an attempt to demonstrate that the light from the abdomen of a glow-worm (the larval stage of a beetle) was pale green due to its X-ray content.[130] Leo Brett saw by the odic light of X-rays.

Mason didn't equate the light Leo saw with X-ray tube glow. Nor did Edward Macomb Duff and Thomas Gilchrist Allen when they quoted Mason in their 1902 attempt to use psychic research to account for miracles narrated in Christian scripture.[131] The etheric double of a person is visible as an aura, they contend, citing Reich-enbach's experiments with clairvoyants. Mason's passage on Leo Brett's pale green light they give in italics. This special color of the light merges Leo Brett's greater than X-ray vision with odic force and the light that suffuses miraculous events in the Gospel.

Dr. John D. Quackenbos compared Leo's X-ray vision to clair-voyance achieved through hypnosis.[132] Quackenbos was a physician

---

[129] Mason (1901: 255n)

[130] The Rays of the Glow-worm, *Electrical Review* 43,1097 December 2, 1898: 835.

[131] Duff and Allen (1902: 146).

[132] Quackenbos (1908: 119-22).

and professor emeritus of rhetoric at Columbia College whose early 1900s adventures (or ventures) in hypnotism and soul recall were chronicled by newspapers nationwide. He made the borderland between science and spiritualism into theater.

He demonstrated his trance cures of physical and moral ills, such as criminality and alcoholism. He "saved" an actress from stage fright by putting her into trance, and held soul-traveling parties with the assistance of a clairvoyant. As "Leader in World of Hypnotism" he was "credited with having recalled to life a young girl only a few minutes dead." He himself made no such claim, but did relate the girl's story of the world beyond.[133]

To Quackenbos a trance state could activate penetrating sight and suggestion could direct the sight. Hypnotism was the indispensable source of Leo Brett's powers. Quackenbos was attracted to a Rhode Island girl, Beulah Miller, who achieved X-ray vision without being placed in trance.

Quackenbos followed other psychic researchers seeking instances of Beulah's in-sight. Beulah's mother later told inquirers that her daughter's gift first became noticeable during family card games when she always seemed to know the cards in another player's hand. Beulah first was brought to public attention in 1913 when she was ten years old by the minister of her congregation, Dr. Watjen, who wrote an article about her in a church bulletin. He was followed by a local judge and a physician who performed simple tests, concealing coins and playing cards for her to guess. The word of these prestigious men in turn brought Dr. James H. Hyslop to Beulah's home. Hyslop was on a hunt for evidence.

---

[133] Quackenbos Resigned as Head of Rhetoric Department, *New York Times*, March 27, 1893; Hypnotism is a Cure-All, *New York Times*, April 30, 1899; Spell of Hypnotism Saved Actress, *New York Times*, December 20, 1901; Gives Soul-Traveling Party, *New York Times*, March 3, 1906; Leader in World of Hypnotism, *Bisbee Daily Review*, September 20, 1906: 1.

> Prof. James H. Hyslop, secretary of the American Society for Psychical Research, said recently: "The whole subject of psychical research, covering all the unusual phenomena of mind, promises to give a meaning to the cosmos which is not dreamed of in physical science. Of course, the latter has opened up a vast universe of occult material forces, such as wireless telegraphy, the x-ray, the n-ray, radioactive phenomena, and ions and electrons, all of which make it but a slight step to believe in the possibility of spirits. The only question now is to hunt for the evidence, and if that is obtainable the methods of psychical research ought to be able to produce it.

Hyslop was an academic philosopher who followed European trends in psychical research and helped found an American Society of Psychical Research to match the aims of its English forerunner. In his numerous writings he advocated integration of science, religion and "supernormal research."[134]

Hyslop named Beulah Miller's insight "X-ray vision," and was followed in short order to the Miller residence and in this designation by Quackenbos.[135]

> Beulah Miller undoubtedly possesses x-ray vision, or power to see through opaque bodies-supernormal sight. Her inner consciousness has the gift of perceiving outward things without the aid of organs of sense. So there is justification for a belief in medical clairvoyance or the power of diagnosing the state of a person from direct contact with the eye endued with extracorporeal activity-x-ray vision at short range.

Beulah's ultimately correct ascertainment of objects which someone else had hidden from her view convinced Quackenbos that she saw

---

[134] Sayings and Doings, *Homiletic Review* 54 (1907: 33).

[135] Quackenbos (1916: 214)

through barriers but not through the object seen. She could not perform her feats while blindfolded or in the absence of her mother or a sibling. This was not problematic for Quackenbos, who conceived of Beulah's vision as an extension of her corporeality, as an X-ray beam is an extension of the X-ray tube. Quackenbos was thus relieved of any need to detect genuine X-rays in the vicinity of Beulah's activities.

Beulah's reputation for supernormal sight conveyed in newspaper and periodical stories reached across the Atlantic but not across the United States. It did reach Harvard University where Hugo Münsterberg was head of the psychology laboratory. The German physiological psychologist and physician had been recruited by William James to teach and conduct research in the recently founded psychology department. Münsterberg sought evidence of the correspondence between the bodily processes and brain processes observed systematically and measurably in behavior. He made three visits to Beulah Miller at her home in Warren, Rhode Island, each time conducting tests to quantify the nature of her vision.

At no time did Münsterberg signal acceptance or even serious consideration of the X-ray vision theory. He tested Beulah by placing sequences of three letters on cardboard squares in a box lid out of her view. It took her at most three guesses to determine which letters had been selected. This and other tests convinced Münsterberg that Beulah was not a fraud. He rejected the assertion that her vision was supernormal. "Beulah No Marvel," the headline in a *New York Times* article flung out, "Harvard Professor Scouts Idea That Rhode Island Girl Has X-Ray Vision."[136]

The newspaper coverage of Beulah Miller was not as widespread or extensive as Quackenbos's previous forays. It was confined to New York and Boston papers, which treated it as a contest between the two researchers over her X-ray vision. Münsterberg's article on Beulah in *The Metropolitan* magazine, extended into an article titled "Thought Transference" in a collection of essays, *Psychology and*

---

[136] *New York Times* April 23, 1913.

Social Sanity (1914), continued his plain dismissal by silence of both Quackenbos and the X-ray vision theory.[137]

In a thorough review of his own tests of Beulah and the conclusions others had reached through less rigorous testing, he proposed that Beulah knew the thoughts of others "by subconscious reading of unintended signs." Instead of having a special radiesthetic ability of the eyes, she picked up on almost imperceptible gestures of her mother, siblings and others who had in mind the information she remarkably, but not miraculously, perceived. The final lines of the article regretted the effects of this reputation on Beulah herself but even more "the behavior of this nation-wide public which chases her into the swamps of fraud."

The implications of this final word did not escape Quackenbos, whose reply came in the form of his own review of the Beulah Miller experiences in 1916. He alluded to Beulah's guileless innocence and her simple family background. Beulah liked him but did not like the Harvard psychologist: she did not perform well for those she didn't like. He recounted the many examples of her X-ray vision in action.

Münsterberg's suspicion of her inability to see through barriers while away from home or in the absence of her mother was answered by a test Quackenbos performed with her in his Manhattan office with only her sister present. He grasped one of three bills hidden in his pocket and she guessed the correct denomination every time. He did not provide details of the experiment, as Münsterberg would have asked. Münsterberg, familiar with the currents of European neurology, had made the subconscious the center of his analysis of Beulah's ability, and challenged Quackenbos on home ground. In response Quackenbos titled his book *Body and Spirit: An Inquiry into the Subconscious*.

There is no earthly way of knowing if there would have been a rejoinder from Münsterberg, who died in 1916, nor were there subsequent statements by Quackenbos. Further reported tests of Beulah Miller were not in his favor. As part of his 1916 inquiry "What is There in the Occult?" journalist Bailey Millard wrote in

---

[137] Münsterberg (1913) and (1914: 144-77).

the caption of a photograph of Beulah Miller that James D. Hyslop declared her "the girl with the X-ray eyes" but E.L. Kellogg, President of the Metropolitan Psychical Research Society of New York branded her a "fake."[138]

17. Beulah Miller with companion. Fn138

The lack of unanimity even among Quackenbos's fellow psychical researchers in New York dissolved into the distractions of the coming war in Europe. In 1920 it appeared to some looking back upon the Miller case that Hyslop, Quackenbos and Münsterberg had been taken in by a deception as transparent as that practiced by the Fox sisters in the 1830s: Beulah's mother signaled the answers to her by a code of tappings.[139] In the long run Quackenbos was singled out as the laughingstock of the case.[140] Münsterberg, a pioneer in applied psychology, had been on the right track. He permanently set aside

---

[138] Millard (1916: 634)

[139] Spiritualism and Spiritasters, *American Medicine* 26 (1920: 469-71).

[140] Rinn (1950: 327).

"X-ray eyes" from serious scientific consideration. That scientific rejection only gave it greater traction in popular culture.

11

PERFORMING X-RAY EYES

The "non-professional medium" Elizabeth d'Espérance (Elizabeth Hope, 1855-1919) was able to read letters still enclosed in as many as seven envelopes, and in languages she did not know, such as Swedish.[141] She didn't translate the Swedish letter; she spelled out the words for others present during the experiment to write down.

There always were others present during her experiments. She had spent her childhood alone among shadow people and was scolded by her mother when she attempted to tell others about them. She spent her adulthood communicating with them, drawing them in the dark and materializing them for the camera. People sent her letters to be read through the envelope and especially if the writer were an influential person whom she wanted to convert to spiritualism she looked into the envelope and read the letter. Reading letters she had never seen before through barriers was her way to assert social communication overcoming barriers.

d'Espérance in 1897 published her autobiography written over the previous decades, a chronicle of her growth toward spiritualism and her growing ability to perform her mediumship. She subscribed to a theory of the "subliminal self" that her body housed an unknown being in contact with others both embodied and spiritual.[142] She ceased her mediumship when a catastrophic failure seemed to expose her as a fraud. She had established a standard of performance which could not survive a break in her audience's support based on acceptance of "her" spirits.

---

[141] d'Espérance (1897: 217-20)
[142] Owen (2004: 285n67)

Seeing into envelopes was only one aspect of d'Espérance's mediumship. It verified her ability to exercise her senses where spirits move. The spirits didn't convey the words to her. She could read them on a level of sight more fluid than the mundane. This was Mason's interpretation of Leo Brett's X-ray eyes, and Quackenbos's interpretation of Beulah Miller's. Münsterberg ignored X-rays and rendered Beulah Miller's perception as an unconscious ability to read cues from her relatives. He stopped short of accusing her of fraud, and considered her a victim of a mob demanding supernatural explanations.

Guy Fenley. Leo Brett and Madame Endor were practitioners of dowsing and medical clairvoyance starting up under the X-ray brand. d'Espérance and Beulah Miller were performers of spirit mediumship and clairvoyance using the X-ray brand. When a career was beginning in the presence of X-rays they were a useful means of making an otherwise run of the mill clairvoyance skill seem new and exciting. With time X-ray eyes became a performance in themselves.

Rayon made the marvelous being of Elfa known to the *New York Herald* and other newspapers in 1894, when Elfa was sixteen years old. Three years later a long report by an unnamed woman journalist on a session with Elfa apparently at her rooms in Chicago appeared in several periodicals.[143]

Rayon made it known that Elfa was a sickly girl living in Mount Clemens, Michigan before he began to align her with the zones of magnetism. Elfa, the name Rayon bestowed on her, could enter into what he called a "psychoma." In that sleep-like state-Rayon vigorously denied that she was hypnotized-her pupils dilated, her beauty all the more aglow, she looked into people and located the source of their ills. She listed the contents of a closed purse, including the dates on coins and the denominations of bills. She left her body at will and traveled to other parts of the city, including the inner sanctum of a secret society known only to initiates, and to all

---

[143] Pierces All Space, *Wichita Daily Eagle* December 29, 1897: 8,6
19 (1898: 383-84)

parts of the world. While awake Elfa was a dream of loveliness taking walks with her loyal mastiff Juno at her side.

Elfa "searches the human body with more acuteness, more thoroughness than an X-ray. There are no shadows in her mind." Elfa "describes even the color of thought, and the aura is as plain to her as a picture on a wall." Unlike the dim shadows of the radiograph, Elfa's picture is full of light and color. "At least, so she says," the reporter volunteers.

Elfa does not claim X-ray vision before it is dismissed, like some versions of Leo Brett, to emphasize her superior color-reading skills. Elfa combines several types of clairvoyance, medical clairvoyance and penetrant sight with out of body experiences. She resembles 19th century fasting girls like "the sleeping girl of Brooklyn, Mollie Fancher"[144] in her clairvoyance and out of body travel, but under Rayon's guidance she has left the sickly, bedridden life behind.

Elfa's expanding fame as a clairvoyant intersected with what others were advertising as X-ray eyes. Rayon, writing as the "guide of Elfa" prophesied a revelation.[145]

> The day is not far distant when the public will know that hypnotism is but another makeshift of science under which it attempts to hide its ignorance of the higher human forces. And the same may be said of the "X-ray" effort, which is principally a clumsy attempt to accomplish, by mechanical means, what is done in the very rudiments of that psychic training in which induced somnambulism is the principal factor.

The "X-ray" effort is the attempt to substitute X-ray visions produced by tubes and screens for out of body travel that can pass through all barriers and over all distances. X-rays and radiation were not the given and final truth boasted by science, but another

---

[144] Walsh (n.d.).

[145] Rayon (1897: 134)

failure to achieve by science what psychic discipline already had accomplished. X-rays were the fraud, trying to reproduce the gift of clairvoyance by means of a machine. The promise of television and neurovision was as nothing.

The headline writers of the *New York World* had not consulted Rayon when they led off the article about Elfa with "The Most Marvelous Girl of the Century! X-Ray Eyes!"[146] The editor of the journal *The Theosophist* considered it sensationalist as well, but then handed Rayon's own aspirations a blow by dismissing the account of Elfa as "simply the record of a case in which clairvoyance has been developed by human magnetism, and is not new or strange to those who are familiar with magnetism." The "X-Ray Eyes" banner is just an attempt to boost another example of a well-known procedure. Rayon is grouped with all the other upstart psychic guides trying to grab attention.

Both Rayon and *The Theosophist* cast the resort to X-rays as an ephemeral promotion of a psychic power not susceptible to quick blasts. After the attempt to make Elfa into a medium able to materialize spirits for the camera both Rayon and Elfa faded from view.[147]

X-ray eyes remained a viable framework for promoters of clairvoyants. The community and the parlor stage where dowsers, clairvoyants and mediums gave their readings gave way to the theater stage, where X-ray eyes became an act. Occasionally a notice like the following appeared in the entertainment section of metropolitan newspapers.[148]

> Remarkable feats are performed by Shireen, known as "The Girl With the X-Ray Eyes." After double and triple bindings are placed over her eyes, she is able to describe all sorts of objects placed before her.

---

[146] Magnetic Sleep and Clairvoyance, *The Theosophist* 19 (1898): 383-84

[147] Rayon (1900)

[148] Nixon's Grand, *Evening Public Ledger* (Philadelphia) October 25, 1921: 18

There is nothing new about a blindfolded performer reading texts or identifying persons and things seen by others. The change is in billing her as "The Girl With the X-Ray Eyes" among other acts. Previously these performances took place in the parlors of private homes and were treated as scientific experiments. Now the audience was larger and paid at the door expecting to be entertained by a set of unrelated acts. "Variety" was the name of the milieu.

In becoming an element in admitted entertainment X-ray eyes didn't have to be explained except by patter, which kept the attention of the audience where the performer wanted. Having moved from the realm of painstaking scientific controls on the materials and conditions of the experiment, the person with the X-ray eyes avoided scrutiny where the illusion was made, as in all stage magic.

The audience members knew it was a trick. The entertainment value lay in the trick not being obvious. When X-ray eyes moved into the performance arena away from scientific controls and psychophysical explanations, they became more subject to the criteria of entertainment, satisfying a wish to be pleased by the magical art but not deceived by fraud. Demonstrating that the blindfold was tight and impenetrable replaced the pretense of supernormal gifts, psychic powers and spirit assistance.

The status of X-ray eyes as an act was clarified by Harry Houdini's 1924 intervention in the performance of Joaquin Maria Argamasilla. The 19 year old Spaniard was a son of the Marquès de Santa Cara, and as a member of the nobility disdained performing on stage. Instead he kept to the salon demonstrations and scientific experiments of the previous generation. Together with his younger brother he made the rounds of psychic investigators in Spain and France, by his own account impressing the physiologist Charles Richet with the precision of his readings through metal containers.

Richet did not accept spiritism, that there were communications from the dead, but he did make rigorous experiments with psychics that convinced him such phenomena as clear-sight and telekinesis, for which he coined the collective term "metapsychique," were genuine.[149] Studies were required to measure the force of trans-

---

[149] Richet (1933: 143-56)

cendent sight and hearing. Richet did not list Argamasilla among those he tested. The young man's claimed ability certainly fell within the realm of the possible for the physiologist.

In 1923 newspapers in America reflected notices in Spanish newspapers of the Argamasilla brothers.[150] The following year Joaquin Maria appeared in New York with his manager "Mr. Davis of Brazil."[151] At the Hotel Pennsylvania in Manhattan he sat in a window seat in bright light and read the name "E. Hendricksen" through a silver box, "Munehira" (missing one letter) through an iron box with a zinc lid, and told the time registered on the dial of a gold hunting watch through its closed lid.

Houdini, standing among those gathered for this performance, issued a challenge. He wagered on 5:25 odds that he could duplicate Argamasilla's feats. Mr. Davis accused Houdini of maliciously charging Argamasilla with deception, and left the room with his charge. After their departure Houdini gave his own reading of the closed watch, missing the correct time by one hour.

Two days later the parties met again but were unable to agree upon the conditions for a match between the two protagonists, whose partisans traded accusations of trickery.[152] Houdini did not reproduce Argamasilla's readings, and the one attempt by Argamasilla to read through a box ended in failure. Houdini had placed in the box a real estate advertisement with print so small that few present could read it without eyeglasses. Argamasilla said the day was too dark, that he needed "super-light" to see through the metal. As darkness fell, the test ended.

"This phenomenal mystifier essayed to perform or accomplish the impossible," Houdini wrote in his own account of the encounter," he makes claim to a power of supernormal vision, X-ray eyes and a penetrating brain."[153] It was Houdini's mission to expose any pretense to supernormal powers. He positioned himself near enough to Argamasilla as he held the box to his head that he noticed

---

[150] Gifted with X-Ray Eyes, *New York Times* March 3, 1923
[151] Challenges Super-Sight, *New York Times* May 7, 1924: 3
[152] Clashes with Houdini, *New York Times* May 9, 1924: 22
[153] Houdini (1924)

a slight give in the lid fastening that allowed the light to stream in as he passed it before his eyes. This was the reason Argasamilla would only use his own boxes and would only read in "superlight". Houdini later included boxes like those Argamasilla displayed in his own act but to demonstrate that it was construction, sleight of hand and timing, not special powers that enabled to young man to read through metal.[154]

In early 1924 Houdini was making a successful enterprise of exposing the tricks of mediums and others pretending to greater than human powers. He created an encompassing act that laid bare their machinations. He had just appeared in a film, *The Man from Beyond*, and had been booked by the Albee-Coit organization to give a series of lectures at theatres in western and southern states. He published pamphlets on his exposures-Argamasilla had second place in a pamphlet on a Boston medium named "Margery"- and he soon would publish a book, *A Magician among the Spirits* that detailed his demystifying career up to that point.

Argamasilla relied upon the respect and distance accorded his noble status in a gathering of men to carry off his illusions. His insistence on bright light to illuminate the interior of the box, an ornate object to distract the viewers, came closest to substituting actual light for the X-rays other remote readers claimed were at work. As a photo of the encounter shows, Houdini did not keep his distance, and he equaled the grandeur of his rival's props by offering a 5:25 rather than a 1:5 return on bets. He stage timing and familiarity with metal fastenings made him able to play the role of both participant and observer.

Argamasilla retired to Spain where he occupied himself with writing historical novels of the bygone glory days of the Spanish Empire, and a study of somnambulism. Enrique Márquez showed in 1991 that there were those in Spain who still nursed the belief that Houdini did not make Argamasilla out to be a fraud.[155] Whether out of ignorance or due to a misguided nationalism, he could not say.

---

[154] Houdini (1953: 256-57). Whether as the result of a typo or by design Houdini refers to the Spaniard in this printed lecture as "Argamasilli".
[155] Márquez (1991).

18. Houdini confronts Argamasilla. Fn155

The Argamasilla-Houdini encounter marked the final turn to performance for X-ray eyes. If they were an occult skill, they were perfected among other types of clairvoyance attained as a stage in advanced meditation and not subject to the critique of a Houdini. Swami Panchadasi had years earlier written in his course of advanced lessons in clairvoyance that for an adept X-ray clairvoyance is "as simply natural as is the X-ray."[156] No real X-rays are involved. Seeing through opaque bodies with the eyes is no more complicated than the passage of an X-ray beam through the same bodies.

Magic trick manuals offered step by step instructions how to fake clairvoyance. One trick, entitled "X-ray eyes" set up the deck so it appeared that the card dealer was looking *into* the cards held by an audience member, an old trick with a new name.[157] By the 1930s X-ray eyes were a skill to be learned and a party trick to be per-

---

[156] Panchadasi (1916: 145)
[157] Fischer, Mussey and Oursler (1931: 132)

formed. The name was applied in a spirit of willful suspension of proximate scrutiny by onlookers in order to convey an innocent power of performance.

Kuda (pron. "Q-da") Bux volunteered to have his fire walking endurance tested by the psychical investigator Harry Price.[158] In September, 1935 the native Kashmiri stepped across a pit of coals formed of burning wood, newspaper and paraffin without visible pain or damage to his feet. An English volunteer did not fare so well. Price photographed, filmed and measured the walk and the walker, and could detect nothing that would have insulated Kuda Bux's feet. They were soft and measurably cooler afterward than before he stepped across the pit.

Authorities did not permit Kuda Bux to set up fire pits inside theatres, so he continued his stage career as a blindfolded performer. His eyes were sealed with cotton pads and his head was wrapped with bandages, leaving only an opening for his nose. He read or traced what people wrote on a blackboard, imitated hand and body gestures, and on one occasion rode a bicycle through London traffic while blindfolded. A typescript report in the University of London's Senate House Papers examines Kuda Bux's claim to see with his nostrils.[159]

Harry Price did not refer to Kuda Bux's X-ray eyes in his accounts of the man's performances. After Kuda Bux made the transition to the United States in 1938, introducing himself with an outdoor fire walk for a broadcast of the Ripley's Believe It or Not radio program, he concentrated on his eyeless sight stage performances, began to call himself "The Man with X-Ray Eyes" and was reported by newspapers and periodicals under that name.

Joseph Dunninger, a professional magician on the investigating committee of the Universal Council for Psychic Research, related Kuda Bux's firewalking to a "miracle" staged by Shinto priests who fired up the periphery of a circle and walked across the cooled-down center.[160] This furthered the Kashmiri's association with

---

[158] Price (1939: 250-52).

[159] Pear (1935).

[160] Firewalker Trick Bared by Expert, *New York Times* September 19, 1935

Asian exotics, but didn't give away the secret. Others who walked the same course as Kuda Bux did were badly burned, and expert observers present didn't find that deceit was practiced.

Besides the blackboard and bicycle performances, his repertory included threading a needle, telling the colors of balloons or fish, reading books and magazines, and hitting a target with a rifle shot,

19. Kuda Bux reads a newspaper. Fn161

all with his eyes sealed and his head wrapped in cloth like a collapsed turban, usually done by members of the audience on his instructions. West Peterson relates that he removed his drivers card from his wallet and asked the blindfolded Kuda Bux to read it in the wallet. Kuda Bux said he would if he could see it.[161]

In 1948 he responded to requests by viewers of the NBC show *I'd Like to See...* doing what they wanted to see him do blindfolded.[162] A fifteen minute Saturday evening show, *Kuda Bux, Hindu Mystic,* was broadcast by CBS until June, 1950. The following month a review of his vaudeville act as part of a program at The Palace in Manhattan did not mention any blindfolded acts.[163] "Inadequate salesmanship" marred his illusion of serving the audience with glasses of water drawn from a tap outlined on paper. At the beginning of a filmed episode of blindfolded reading of blackboard writing from one of his television appearances he introduces himself as the "man with X-ray eyes."[164]

Racecar driver Tommy Irwin, who was active on the NASCAR National Circuit from 1958 to 1963, recalled Kuda Bux driving cars around the tracks and through obstacle courses with a doughball placed over each eye.[165] To the amazement of Irwin and other drivers he burst balloons with shots from a .22 rifle. Women ran away when he appeared because they dreaded his stripping gaze.

Kuda Bux attributed his fire walking and his special sight to yogic discipline, and gave a few demonstrations of his ability to convey the discipline to other people. He considered himself, and was accepted as, a stage magician but he never disclosed secrets of his tricks, maintaining his "Oriental mystique" throughout a long career.

He placed advertisements in New York newspapers inviting the public to free demonstrations of X-ray vision. The small crowds welcomed to these modest residential apartment venues witnessed Kuda Bux performing his head-wrapped readings and color identification. Vincent Daczynski went to two of these shows and did not

---

[161] Peterson (1949)

[162] Huff (2006: 15)

[163] Bill Smith, Palace, New York, *The Billboard* July 8, 1950: 42.

[164] Video of Kuda Bux TV appearance. www.geniimagazine.com/wiki/index.php?title.Kuda Bux

[165] Strosnider (2002: 85).

discern trickery.[166] Kuda Bux told the gatherings that he would teach his technique for free to anyone willing. Daczynski was not willing to invest the hours staring at a burning candle that the illusionist said would awaken his powers.

Kuda Bux's assimilation of the X-ray eyes label was part of his general transition from an imported Indian fakir to a performing magician with powers explicable only as the result of religious discipline. At his first firewalking test for Harry Price he insisted on certain provisions due to his Muslim faith, but did not repeat that identification thereafter. It seemed sufficient for him to be known as a "Hindu" when a religious label was used. His widely publicized marvels renewed the X-ray eyes image and shifted it from psychic research to the performance arena, while maintaining a professionalism that insulated him from would-be Houdinis.

He was the last stage performer to make a profession out of X-ray eyes feats, or to claim that label.

Only later in a 1937 article on a thirteen year-old Glendale, California boy who could play ping-pong, drive a car, read, recognize cards pulled from a deck and imitate the actions of others all while eye-patched and blindfolded, did the reporter call them "X-ray eye stunts."[167] The boy was discovered by a doctor with training in London, and the narrative gravitated toward the psychic research model, with a distant memory of the boyish Leo Brett. To the reporter he seemed the least likely of his family to be psychic, when seen next to his siblings and his "pale, dark-eyed mother." Later the psychic investigator J.B. Rhine caught Pat peering down his own nose through a gap in the blindfold.

Claimants to X-rays eyes today are performers who incorporate tests by professional skeptics into their displays, if they can gain enough notice to be tested in the first place. The media and the skeptics must see sufficient benefit to themselves in staging an individual spectacle of "paranormal" abilities. The abilities are depar-

---

[166] Daczynski (2004)

[167] Pat Marquis of California Can See Without His Eyes, *Life* April 19, 1937: 57-59.

tures by measurable degrees from normal perceptivity, or they are available to anyone who uses recording equipment.

A corollary of this is televised "ghost hunting," which follows "normal" individuals intrepidly entering upon premises reputed to be haunted because of known events or suspicions arising from the prior use of the place. Those with X-ray eyes are a small and infrequent component of televised paranormal investigations, which dispense with the skeptics to provide seemingly unscripted dramas with non-professional participants in the manner of reality television.

Those endowed with X-rays eyes appear outside of the EuroAmericanAsian media ambit, and manage a tentative entrance before disappearing again. Or they find enough benefit from modest national renown.

Natasha Demkina was introduced to English audiences by television appearances during a tour first sponsored by the English tabloid *The Sun* in 2004.[168] Seventeen years old at the time, she was a native of Saransk, a Russian-majority city of semi-autonomous Mordvinia province. Demkina came to notice as the result of an article in *Pravda*, and demonstrations of her talent on Russian television. On English television she looked into the body of a talk-show hostess and located injuries she had received in a car accident. Demkina recounted her first insight: at the age of 10 she saw the interior of her mother's body in colors. Since then she has been able to engage her medical vision at will for a fraction of a second, before returning to normal sight.

A *Discovery Channel* documentary on Demkina[169] begins with footage of the dilapidated Soviet-era apartment block where she lives with her mother in the equally depressed city. The stairway to her their apartment is crowded with petitioners who arrive to have their ills interpreted. Their poverty and frustration with the inadequate local health care services provided by the state lead them to Demkina, who at the beginning did not charge a fee but later began to collect modest amounts.

---

[168] Sample (2004)
[169] Body Shock: The Girl with X-ray Eyes (2004)

Russians have long been patrons of psychic diagnosticians and healers, and television advertisements for several appear on the screen. A journalist, Igor Monichev, tells of his conversion from skeptic to believer in Demkina's powers. A man who lay in a hospital languishing with tuberculosis persistent in spite of treatment received a drawing from Demkina of what she saw in him. A doctor at the city's hospital compared the drawing to the tissue sample she observed under the microscope, and pronounced the man's condition sarcoidosis granulosis. Demkina says in English that she hopes to earn enough money to move with her mother to Moscow and attend medical college there. Most of all, she is quoted and says herself, she wants the people to accept her gift.

The results of the tests prepared for her by members of CSICOP (Committee for the Scientific Investigation of Claims of the Paranormal) in the first battery had Demkina pinpointing afflicted areas of the body of subjects. In the second televised trial she was given a list of conditions (written in English and Russian) and asked to match each of 7 people with the correct condition. She managed to identify 4 of them, one short of the 5 CSICOP stipulated as the threshold of genuine X-ray vision. A number of bloggers and columnists took Demkina's side, supporting her assertion that the test was fixed, and that 5 correct matches was an arbitrary criterion. A set of tests in America ended similarly, with Demkina's authenticity in doubt but defended. In Japan she fared slightly better.[170]

Demkina was titled "the girl with X-ray eyes" in Russian, English, American and Japanese media. She herself did not use the phrase or explain her readings in terms of X-rays. In later interviews her account of the body evolved. The afflicted organs inside patients radiated light, and had an appearance similar to illustrations in anatomy texts. Demkina saw best by day, and was unable to visualize her own organs.[171]

---

[170] Skolnick (2005); Russian X-ray Girl Thrills Japanese Scientists with her Remarkable Gift, *Pravda*, April 20, 2005 [online]

[171] Russian X-ray Girl Natasha Demkina Still Uses her Gift to Help the Common People, *Pravda*, May 29, 2008 [online]

Demkina did enroll in medical college and move to Moscow with her mother as she hoped. Philip Warnell made a film of Demkina reading the interior of his body with theremin accompaniment while they both stood facing each other on the floor of a gymnasium.[172] In his essay "Pregnable of Eye: X-rays, Vision and Magic" written to accompany the film, the cultural phenomenologist Steven Connor makes no reference to Demkina.

Connor finds that the key modern experience is not in the constitution and dissolution of the subject, but in being permeated. The magical thinking of X-rays, a vision that can penetrate everything, cannot see into the power of its own fantasy. Demkina, however, has confined herself to looking helpfully into the bodies of the poor, of television personalities and performance artists. The X-rays are provided by others.

A young woman from a village near Varanasi, India, Ranjana Agrawal, won India's National Child Award for Exceptional Achievement in 2002 for her ability to describe people standing before her while her eyes are obscured by a blindfold. Reflecting Kuda Bux, who never performed in India, she appeared on national television performing her feats. Her blindfold is itself transparent: the gap through which she can look out is visible to an inquirer.[173] Unlike Kuda Bux, who never was "found out" she simply passed across the performance stage and vanished, to be replaced by other miracles.

---

[172] The Girl with the X-ray Eyes (2008). The publication of a pamphlet, Warnell (2008) accompanied the first showings of the film. It contains a version of Connor's essay.

[173] Nayak (2010)

12

CHIEF BENDER'S EYES

Before Pat Marquis (1937) or Kuda Bux (1949) was presented in this way, Chief Bender's X-ray eyes (1911) were shown gazing through a slit-like a break in the printed surface of the page.[174]

Charles Albert Bender was a major league baseball pitcher of German-Chippewa descent. He parried the label "Chief" bestowed on him and other Indians in baseball but he always signed his autographs "Charley Bender." He pitched a no-hitter against an opposing team in 1910, the year before he was the main force in the World Series victory of the Philadelphia Athletics. He was elected to the Baseball Hall of Fame in 1953, the year before he died. For a long time he was the only American Indian player so honored.

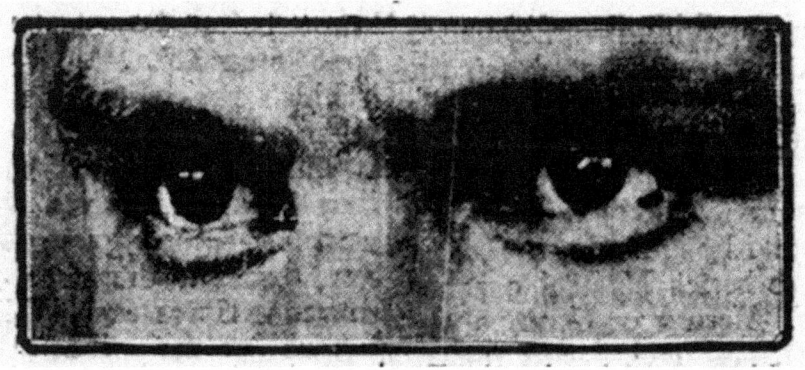

20. Chief Bender's Eyes. Fn174

[174] Tip Wright, Chief Bender's X-Ray Eyes Beat Giants, *The Day Book* (Chicago) November 1, 1911

Baseball journalist Tip Wright ascribed X-ray eyes to Bender because of his sign stealing. Today sign stealing is the disliked (but not illegal) practice of figuring out the hand signals given by the catcher to the pitcher and relaying them to the batter, to prepare him for the shape and speed of the pitch. In Bender's heyday sign stealing was the ability to determine the nature of the pitch by minutely interpreting a pitcher's preparations and communicating that information to the batter. Bender was eminently suited to make these calls and to teach other players.[175]

> Bender originated "calling the turn" from the catcher's box. He taught Coombs and Hartzel, both keen-sighted. Between them they have broken many a pitcher's heart. From first and third they watch the pitcher's fingers grasp the ball. They can tell whether he will throw a curve or fast ball as he winds up. A movement of the foot, hand or body informs the batter and he is ready.

Sign stealing was "calling the turn" in Bender's time. It could also be "signal theft" with disapproval creeping in. To maintain a "tipping bureau," a group of players who could read the signals was worthy of a remark by a baseball columnist.[176]

Bender had vision sharp enough follow the details of a pitcher's positioning and relate them to his own preparations for the throw. The few commentators who use the "X-ray eyes" expression to describe Bender's insights also refer to him as a "Chippewa." No one draws explicit conclusions from the Indian identification, which was a constant in writing about him and generally respectful.

In 1911 the Indian wars, and the emergent mythologization of the participants, were still an active enough memory for the sharp-eyed Indian scout to be a shared image. X-ray eyes were not a racial

---

[175] Chief Bender's X-Ray Eyes Beat McGraw's New Yorkers, *Washington Herald* November 5, 1911: 4

[176] J.W. McConnaughy, Indian Dangerous on Coaching Lines as Well as in the Box, *El Paso Herald* October 8, 1913

attribute but a skill to be taught to non-Indian teammates. None of the books on Bender, and for that matter none of the later journalism on Bender, makes mention of his X-ray eyes, though most refer to his keen vision and sign stealing.

Art Fletcher, during his years as a coach for the New York Yankees, was also said to have X-ray eyes for his ability to "denote the slightest move by a pitcher which might telegraph what he was intending to do."[177] Fletcher's life span (1885-1950) coincided with Bender's (1884-1954) but apart from that they bore little resemblance to each other as Major League Baseball players. That the Paleoindian mound complex Cahokia is within the city limits of Fletcher's home town of Collinsville, Illinois has less to do with Fletcher's sign stealing than the fact that Fletcher was third-base coach during the Yankees' streak of World Series wins.

X-ray eyes in this case is a casual phrasing of a kind of attention to the motions of the game. When that attention became sign stealing, X-ray eyes were no longer in view. Yet its attachment to Bender is haunting. There are no photographs of Art Fletcher that catch the look of his eyes as there are of Bender, a dark-eyed look that Wright noted in his original article. Bender truly had eyes that suggested he was looking through something.

Bender had both X-ray vision-the ability to look into the pitcher's play-and X-ray eyes, the dark pupils that gaze intelligently. His on-field presence was complaisant and generous. He taught his skill to teammates and cooperated with them to bring in the victory. He always was shown smiling.

Bender had been preceded in Major League Baseball by Louis Sockalexis, a Penobscot Indian from Oldtown, Maine. Sockalexis hit a home run facing racial taunts from the bleachers and a pitcher who swore he would strike out "the damned Indian" as he played in an exhibition game at the Polo Grounds in New York.

Shortly after that play he was expelled by Notre Dame University because of an incident while he was drunk, and was soon signed by the Cleveland Spiders in the National League. His entire Major League career lasted two years from 1897 to 1899, when he was

---

[177] Steadman (1982: 44)

dropped from the team and returned to Maine where he coached up-and-coming Penobscot players.

In May of that golden year, 1897, a sports writer asked Sockalexis about the shouts directed at him from the stands during games. Sockalexis said that wherever the team played he went through "the same ordeal" and "at the present time I am so used to it that at times I forget to smile at my tormentors, believing it to be part of the game."[178]

Perhaps Bender was familiar with the conditions that inspired that often repeated quote. He maintained a smiling demeanor on the field, but his eyes told a different story.

---

[178] Feitz (2002: 85)

13

THE TRANSPARENT WORLD

Inventions and discoveries bring to the fore social contentions that long existed but were not as explicit as the new terms make them seem. Adopting the X-ray manner of seeing and looking led to disagreements. Those attempting to make parts of the world transparent as might appear to the imagined X-ray eyes but for everyone to see were also in peril of reputation. The transparent world was uncovered long enough for a glance.

The "X-ray skirt" or "X-ray gown" would seem to be just another metaphoric application of the "X-ray" brand bringing into contemporary fashion an existing item of dress. Editorialists were at pains to derive it from earlier fashions. Far from being modish it harked back to a primitive form of female accoutrement, the "flesh-colored gauze chemise over pink fleshings with golden garters, and a fillet of gold in the hair" of an earlier era's *femme sauvage*.[179] One writer compared it to the garb of women visible in Babylonian friezes. Referring it back to presumably looser periods in human history gave the X-ray skirt the quality of a long-term vice, like gambling and prostitution, railed against but always present in a form.

All writing about the X-ray skirt adopted a condemnatory moral tone, often grouping it with other revealing garments also being rehearsed at the same time, the high-slit skirt and the tight skirt, which disclosed the figure at the cost of confining the legs. Tights of cotton or wool did exist, but no one ventured off the variety stage

---

[179] How To-Day's Styles Ran Their Course 100 Years Ago, *New York Times* July 27, 1913; X-Ray Gown Women's Most Primitive Dress, *The Day Book* (Chicago, Illinois) September 23, 1913: 25

wearing them on legs completely exposed. As my wife reminded me when I asked her if she had ever heard of X-ray skirts during a career in the fashion industry, the time of their widest currency was when some households put skirts on piano and table legs lest there be a suggestion of the shapely female leg in the sitting room.

The exposure of women's legs was the primary issue. A vision of their shape was a boundary the public gaze approached and wrenched itself away from with a force displaced to authorities. The X-ray name was not just a metaphoric label making an old challenge in contemporary terms. The legs (and nothing above them) were cast in the shadowgraphic fog of an X-ray plate. Their shadow was offered to those ogling women through imaginary X-ray spectacles in a precocious version of "what you see is what you get."

The assemblage worn by a Boston society lady-"black lace worn over black silk knickerbockers and black silk stockings"-describes the shadow play of the X-ray skirt.[180] A vaudeville performer named Vera Hall met a dare and walked up Pacific Avenue in Tacoma, Washington wearing a "diaphanous skirt."[181]

She told the reporter-photographer that she was humiliated by the stares of the men and would never do it again. She said that it "depends on the legs" whether or not a woman should wear the skirt. It did give greater freedom in walking, and was "cool and sanitary." She recalled a French woman who dined with her husband at the house of her parents, and the upper part of her body politely laid bare as she sat at the table. Turkish women, she added, would "blush to death" at the thought of going bare-faced and unveiled. Vera Hall attributed the anxiety about diaphanous skirts to values that vary from one community to another. The photograph that accompanied the front page article reinforced her statements. She stands erect the shape of her legs scarcely visible within the curtain of her skirt. Directly below the cinched-in waistline of her jacket is

---

[180] This Transparent Gown for Boston, *The Bisbee Daily Review* August 23, 1913: 1

[181] Oh! Naughty! Woman in a Real Diaphanous Skirt Walks Up Pacific Avenue, *TheTacoma Times* September 6, 1913: front

21. Vera Hall in her "diaphanous skirt." Fn181

a dark line that lowers the already grainy contrast of the photograph. The newspaper is making it clear that the thought of the skirt is all that will get through on the front page. Vera Hall's boldness combined with reticence is carried over into the photograph.

A cartoon of the X-ray skirt can have higher resolution.[182] An X-ray hat is among the other fashion features that include a monocle for women and a veil that positions a gleaming glass disk over the eye. X-rayed clothing and radiant eyes, the two aspects of X-ray

---

[182] Single Eyeglass; The Real Thing-Also is the X-Ray Skirt Direct from Paris, *The Day Book* (Chicago, Illinois) July 28, 1913: 27

22. Look Boys-Direct from Paree The New Xray Skirt Fn182

beauty for women. Another X-rayed item of personal wear advertised in that key year of 1913 was the X-ray shoe (for women only).[183] The upper was made of an Irish crochet for evening wear. A street version retained the finest lace for the upper with galoshes of suede kid.

There were no reports of this footwear causing a stir when seen on the street, or forming part of a wardrobe centered around an X-ray skirt. A few beauty contests drew an audience with a promise of women in X-ray skirts.[184] Most of the accounts of women wearing X-ray skirts on the street were monitory in nature. The women

---

[183] X-Ray Shoe, Latest of Fashion's Fads, *El Paso Herald* September 12, 1913: 11

[184] Dolls in Slit Skirts and X-Ray Gowns Vie for Oakland Prizes, *The San Francisco Call* December 4, 1913: 4

wearing them or the men viewing them were required by police to desist.

While Vera Hall walked up Pacific Avenue in Tacoma, the mayor of Portland, Oregon ordered police to arrest any woman appearing in public wearing an X-ray skirt. The reaction of municipal authorities was restrictive and prohibitory. From Los Angeles[185] to Janesville, Wisconsin[186] the police, under orders from elected officials or as individual officers, were decisive in curbing the menace. A policeman standing on the street in Janesville did not accept a woman's protest that her X-ray skirt was the latest fashion in Chicago and commanded her to leave the town within ten minutes or be taken into custody. She quickly left.

The Women's Christian Temperance Union took time from their drive to make Prohibition the law of the land to condemn diaphanous skirts.[187]

Elsewhere a process of negotiation softened restrictions. A woman who had worn an X-ray skirt on the street in Tucson, Arizona and received a sharply denunciatory letter demanded that U.S. authorities arrest the critic for misuse of the mails, and offered to wear her skirt to the trial.[188] A 7-year old girl playfully dressed in an X-ray skirt on the street in Ventura, California, the first such manifestation in that town, defused the conflict with her lack of self-conscious and self-centered cultivation of attention.[189] A man who dropped his dark glasses and "rubbered" [rubber-necked] a passing woman in an X-ray skirt was the one arrested, for fraudulent begging.[190]

---

[185] Police Ban X-Ray Skirt, *The New York Times* August 23, 1913: Front

[186] X-Ray Gown Banished from Janesville, Wisconsin, *The Telegraph-World* July 23, 1914: 11

[187] W.C.T.U. Against Diaphanous Skirts, *The University Missourian* (Columbia, Missouri) October 3, 1913: 1

[188] She Wants U.S. to Jail X-Ray Gown's Berater, *El Paso Herald* September 30, 1913: 6

[189] What Makes Dress Modest, *The Oxnard Daily Courier* October 4, 1913: 2

[190] "Blind" Sees X-Ray Skirt, *The Steuben Farmers Advocate* October 1, 1913: 8

Trying to be equitable (better for business) a dry goods trade newspaper printed a two column pro and con compilation of statements on tight skirts.[191]

TIGHT SKIRTS—PRO AND CON

Every question has two sides.
That includes even the dress problem.
Both sides are stated in these two news items that appeared in the newspapers almost simultaneously:

PRO

"The X-ray and the slit skirt afford freer circulation of air about the body," Dr. Wilbur F. Cannon, of Denver, is quoted as saying. "Air is a tonic and stimulant to the skin Absence of so many underskirts makes less weight upon the hips and may be called a prophylactic measure in kidney troubles. It may save backaches. Absence of the usual amount of clothes necessitates more frequent bathing, thus opening the pores and causing freer perspiration. This means less burden on the kidneys. The latest fashion permits of freer movement of the limbs and conserves energy, and less labor is required in ironing, washing and taking care of clothes."

CON

Following a public statement to the effect that many accidents to women on car steps are the direct result of high heels and hobble skirts, the Pennsylvania Railroad has posted on every trainman's bulletin board along the main line an order that the dress of the woman shall be specially noted and reported whenever such an accident occurs. The order directs that trainmen at all times, as in the past, shall lend passengers assistance, and that whenever a woman falls on or near a train the trainsman who sees it shall jot down the width of skirt and height of heel.

The benefits and detriments of the skirts were according to this conspectus entirely based upon their impact upon the health and safety of the wearer, and the liability of the railroad. Opinions about how women looked and were seen while wearing them were not part of this balance sheet. X-ray transparency turned opaque in the interest of commerce. Those concerns, while immediate, did not trump the spectacle.

There was no brighter sign of the social forces arrayed against each other in the X-ray skirt debate than the case of John Carey's lights. Carey was a merchant with a store near the intersection of

---

[191] Tight Skirts, Pro and Con, *Dry Goods Reporter* 43, 2 (1913): 8

Broadway and 38th street in Manhattan. In August, 1913 he set up four acetylene lamps aimed toward the street in front of his building.[192]

As the police put it before Magistrate Breen of the municipal court, the lights acted "like X-rays" on the clothing of passersby, outlining the body within the dress. This was enough to attract a substantial crowd of men and boys eager to take in the outlines which in turn created a public order problem for the police. Carey defended himself to the police and in court saying that the lights were placed within his property line and were merchandise for sale. Pressed by the police, he voiced his discontent with women being "partly clad" and declared that they (the police) should "change the style." The exasperated magistrate referred the matter to the Board of Health.

An enterprising merchant used a powerful technology (designed for automobiles) to remove the shadings from the X-ray skirt impression and enforce his own sense of propriety in dress. The police might look the other way when a woman wearing an X-ray skirt appeared but they could not ignore the crowd that Carey's exposition caused to form. Blazing lights were a vexation not unknown to urban Boards of Health, though the tenor of this case must have been unusual for them. If Carey's illuminating gesture was meant to attract business it must have reached a point of diminishing returns. He did, however, dramatize the strains of male gazing, female defense and the crescendo of disorder that surrounded the transparent world opening about the X-ray skirt.

That crescendo in New York did not reach the pitch of riot police faced in other cities.[193] At least one newspaper found it useful to restage Carey's gesture locally with some of the names changed.[194] A degree of frustrated acceptance is mirrored in the mid-1914 words of columnist on the unsuccessful efforts of a hus-

---

[192] X-Ray on Skirts Puzzles Judges, *The Washington Herald* August 15, 1913
[193] Sydney, Australia, for instance. The X-Ray Skirt, *Hawera and Normanby Star* October 14, 1913: 5
[194] Aids X-Ray Gazers, *The Daily Republican* (Cape Giraudeau, Missouri) August 15, 1913: 4

band to prevent his wife from going out in public wearing an X-ray skirt.[195] The style had been domesticated at least to that degree.

The domestication of a style, the loss of the scandal far from the city, prefaces announcements of its passing. In the Style Bulletin of the Fashion League of America for Summer, 1915 the passing of the X-ray skirt is firmly decreed.[196] Full skirts reinforced and made non-transparent by heavy lining will provide relief to the parties to the recent struggle, if not to those who considered the skirts more healthful. This was the moment when writers about fashion looked back and included the X-ray skirt among by-gone fads, or saw it as an ancient manner of dress resurfacing.

X-ray skirts were attributed to Paris fashion attempting to enter America. Circumstantial accounts of women being seen at the race-track near Paris dressed in the "latest in transparencies" were captions for the usual grainy photographs.[197] When a woman wore an X-ray skirt to the Coney Island amusement park, the applause she received caused her to flee in a taxi.[198] There certainly were diaphanous and tight skirts exhibited in Paris design collections but the French equivalent of "X-ray skirt" was not used to name them. That terminology was confined to American newspapers.

Yves St. Laurent later applied that explicit term to one of his own creations.

As early as 1898 the phrase was used without further description in a Connecticut newspaper,[199] which suggests that it was common knowledge by then. It appears as a casual reference as late as 1932: "Cloudy day-young lady on street in rain Saturday, wearing leopard

---

[195] Clarence L. Cullen, So Wags the World, *The Pittsburgh Press* October 10, 1914: Saturday Page

[196] Passing of X-Ray Skirt, *The Spokesman-Review* (Spokane, Washington) January 23, 1915

[197] X-Ray Gown the Latest in Transparencies, Tights Go With It, *The Evening World* (New York, N.Y.) July 1, 1913: 3

[198] X-Ray Gown at Coney—Men Follow to Applaud—Woman Flees in a Taxi, *El Paso Herald* August 9, 1913

[199] All About the Trimmed Skirt, *The Meriden Journal* July 23, 1898: 5

skin fur jacket—and an X-ray skirt."[200] Fashions cycle and recycle, old fashions are revived, and the common knowledge remained to recognize an X-ray skirt when one appeared. The X-ray skirt reached a peak of interest in 1913 and quickly fell away. If it always was present why did it rise at that time?

Naming a type of skirt after a current technology no longer so marvelous was not enough to create a fashion stir. The X-ray skirt was a visual type arising from newspaper coverage that showed women's legs appearing inside a skirt like an X-ray photograph or the imagined results of viewing with X-ray spectacles. Women advanced to the line of the voyeur's glance and held their ground even if by retreating from view. This was as transparent as the world around them was going to get. The X-ray skirt wearers were not a style type as the flapper was soon to be. The visual type was not the product of an illustrator's designs, like the Gibson Girl, or directly and unmistakably derivative from European models, like so many others. It was an intersection of clothing with the transparent world X-rays made in visions spread by popular media.

Transparent clothing continues to be proposed, manufactured and worn and it might just be designated X-ray clothing.

Painters and sculptors during the early twentieth century did not need to access X-rays to reveal the human body beneath clothing if they chose to do so. They did incorporate into their works outlined images of body interiors as disclosed by X-rays, though this might also have been inspired by "primitive" cutaway drawings of animal viscera known from native peoples of America, Africa and Australia. When artists drew upon X-rays it was to see the world in X-ray light, in shining, oblique, crystalline colors filling the entire canvas, or rendering the members purely structural as a skeleton of everything. Frantisek Kupka titled some paintings after their X-ray source, and Naum Gabo made metal and glass models of things seem amid radiation more than light.

---

[200] You Could See and Hear These Things Too, *The Chehalis Bee-Nugget* (Chehalis, Washington) December 9, 1932: Front

Alfred Hitchcock did not cite X-rays when he showed the shoes of a lodger heard pacing overhead by the landlord in the apartment below in his 1926 silent film, *The Lodger.*

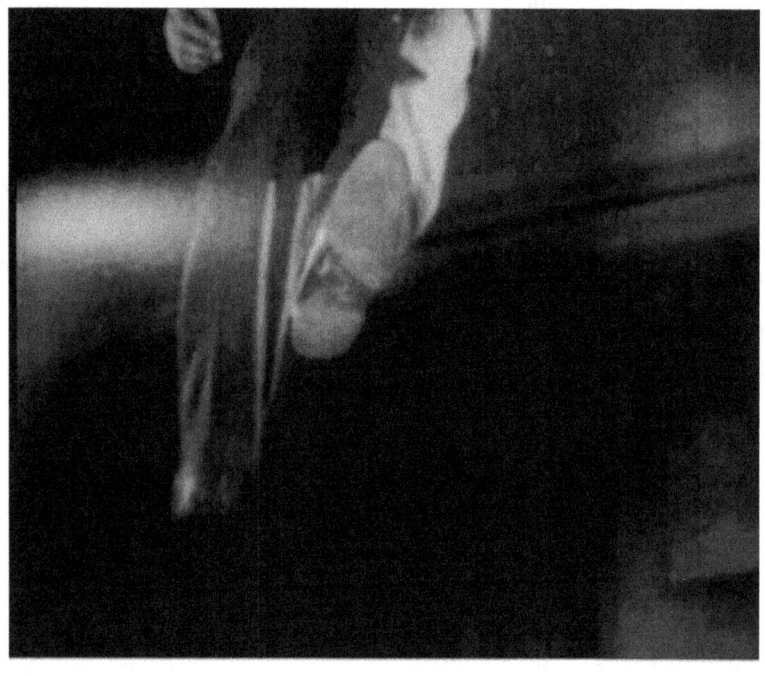

23. Still from *The Lodger* (1926), Hitchcock

In a film he planned between 1926 and 1930, Sergei Eisenstein did cite X-rays.

The son of an architect, and an architecture student himself, Eisenstein approached the making of films as an act of construction leading to revolutionary deconstruction. His *Glass House* script would seem to draw upon other architectural projects, realized and not, to build a residence of transparent material. There were as many theories behind these projects as there were architects sketching them. Eisenstein pictured his filmed building amid X-rays.[201]

---

[201] Eisenstein (2009) which compiles the versions of Eisenstein's script together with a critical essay.

> The transparent building should look like a person under Roentgen rays. The sole opaque object in the glass house, the elevator (a black box with lights like gloomy, all-seeing eyes) looks like a backbone or a key in the pocket [of this X-rayed figure].

This body/building is inhabited by people who themselves have varying abilities to see each other through the transparent walls. A husband cannot see his wife's lover; the rich cannot see the poor. The only person with a universal vision on the level of the building itself is the Poet, whose words no one believes. When the inhabitants lose their bourgeois perspective and see through the walls and ceilings they witness a suicide and a murder.

Eisenstein did not write his script with the direction of a plot. The building is a state of being that the gloomy elevator travels through. The viewer of the film, endowed with X-ray vision by the projector, cannot help but see through everything with the new light shining.

Eisenstein made sketches of scenes to accompany the text of his treatment. Several show shoes seen through the transparent ceiling as in *The Lodger*. The similarity of that sight is the only resemblance between Eisenstein's tower and Hitchcock's cramped house.[202] The transparency of the building, traversed by the elevator with only the doors opaque, opened the eye to a delirium of events connected with each other by a common structure unfolding simultaneously before the traveling eye. This wasn't simply a transparent dwelling like the glass house architectural models but the insides of a body where different motions went on contiguous to each other influencing each other or not. X-rays make the skin like glass covering the organism.

The cinema by the mid-1920s had established its ability to enter rooms, eavesdrop upon private doings, move rapidly down streets and through landscapes. Technical developments had only magnified the speed and perspective of that passage. The X-ray metaphor

---

[202] Jacobs (2007:74)

wasn't necessary to accomplish that degree of social penetration. Eisenstein's *Glass House* was, as Olga Bulgakowa asserts,[203] similar to another of Eisenstein's projects, The Spherical Book, which would place a set of written essays in multidimensional relations to each other. X-rays did not figure into The Spherical Book. Their invocation did make *Glass House* into a volumetric movie like a body projecting all the masses of its operations without regard for dramatic or literary rules.

Glass House, known by its English title because Eisenstein wrote his treatment in English, was like several of Eisenstein's other film projects: it never saw the light of day. Paramount studios quietly shelved the script that a team of writers had worked on in vain. X-ray cinematography would never be realized in the form Eisenstein planned.

Glass House belonged to the same transparent world as the X-ray skirt. Rather than through clothing the surfaces of the building were the body put into the open. X-rays provided a reason not to see through clothing onto the naked skin. The transparent world is an approach and a dare but not a violation of the social order that permits fashion and filmmaking.

Reflecting on an ideal juror's understanding of X-ray evidence, Lyman P. Wilson, a professor at Cornell University's law school, also used the house/body comparison.[204]

> He must know, though a surprising number of the legal profession apparently do not, that you cannot with the X-ray look into the human body as you might look into a house through an open window, nor is the information it can give you exactly the information you could get regarding the inmates of a house, whose silhouettes are by the light within cast upon translucent window shades.

---

[203] Bulgakowa (2005)
[204] Wilson (1922: 203)

The house viewed through a window (as in another Hitchcock movie, *Rear Window*) does not yield information about the house's interior congruent with the information the fluoroscope yields about the body's interior. The shadowgraph of the residents cast on a window shade is not like the X-ray shadowgraph of the body. Wilson also has stopped short of the X-ray transparent world, which now seems to consist of being approached but stopped short of.

An effort at the transparent world by the agency of X-rays came in a story, "The Man with X-Ray Eyes," authored by science fiction stalwart Edmond Hamilton in 1933[205], when the X-ray skirt was making its final appearance and Eisenstein's *Glass House* remained sketches and a script. A young reporter volunteers to receive an eye treatment from a doctor who has devised a vision enhancement formula that will enable the subject to see through inorganic but not organic material. The formula works, and once the reporter learns to navigate invisible walls and poles, he reads the lips of politicians in their conclaves and is completely disillusioned by their venality. His fiancée is also a disappointment: the man with X-ray eyes lip-reads her telling her mother that he is not what she had hoped for but she will take what she can get. His body is found in the river the next day.

The venerable X-ray eyedrops open the transparent world but only to reveal the walls that already exist, not to open them as the reporter hoped.

---

[205] Also titled "The Man Who Saw Everything" Hamilton (1933) synopsis by Bleiler (1998: 166)

14

AMONG OTHER POWERS

The inverse of the transparent world where all things are X-rayed is the presence of X-ray vision among a suite of extraordinary powers and abilities. Prior to the superheroes being invented during the late 1930s X-ray vision was acquired by an individual for better and for worse, but it was a solitary and unique power. The new superhero powers included a forceful vision that always was named "X-ray vision." Its properties reflected the growing utility and menace of the prospectively harnessed natural forces the superheroes embodied: piercing rays and searing radiation. Superhero X-ray vision was more radiation than eyesight, more assaultive than investigative. The vigilantism of the superheroes was empowered by surveillance capabilities.

Olga Mesmer's mother Margot had to remove the bandages from her eyes to realize she could see through walls.[206] Olga's father, Dr. Hugo Mesmer, had rescued Margot, the threatened heir to the throne of a terrestrial subterranean monarchy of Venusian origin, and then in the manner of scientists during the 1930s, married her and experimented on her with his "soluble X-rays." Once the bandages were off her eyes, the rays from her eyes both revealed to her Hugo's dalliance on the other side of the wall and killed the cad. The revealing and destructive force of X-rays were wedded in the same person. The ability to generate that force she passed on to Olga, whom she abandoned shortly after her birth to return to the underground estate.

---

[206] The Astounding Adventures of Olga Mesmer, the Girl with X-Ray Eyes, *Spicy Mystery Stories*, August, 1937. Also in *High Adventure* 44

Olga, "the girl with the X-ray eyes," also received superhuman strength from her mother, and in turn passed on these powers through a transfusion of her own blood to the dashing Rodney Prescott, whom she rescued from an assassin. Thereafter their adventures became both subterranean and interplanetary. Though the magazine of Olga's August, 1937 debut was titled *Spicy Mystery Stories*, and her appearance did feature loose flowing décolletage and signs of a garter belt, the possible first female superhero remained chaste, fending off the advances of her guardian and a Martian prince, who proposed to her on her ancestral planet of Venus.

The monarchic and absolutist planets of Edgar Rice Burroughs and Alex Raymond (Flash Gordon, debut January, 1934) had a place for Olga. Unlike their heroes John Carter and Flash Gordon, Olga was part-human and part immortal Venusian, and endowed with personal energies originating in scientific experiment.

Olga's employment of X-rays was signaled graphically by a lined burst surrounding a blank on the surface of whatever or whomever she was gazing at.

24. Olga Mesmer gazes through a wall. Fn206

After Olga loses her powers to Rodney Prescott, the two of them visit her father's grave hoping to find the formula for the soluble X-rays that will restore her. Instead they are attacked by monsters.

Olga only lasted a few installments of *Spicy Mystery Stories*, and the identity of the strip's author remains the chief mystery. Olga did establish the force of X-rays as a tool of the hybrid superhero-ine. Underline force.

Compared to Olga's mother's first use of X-ray vision, Super-man's was rather contained. The shady stock brokers Meek and Bronson are visiting a pair of thugs to hire them to kill Homer Ram-sey, Superman disguised as an investor who has purchased all the stock in the Black Gold Oil Well, which the brokers are sure will fail to yield oil.[207] News that the well has come in set the two swin-dlers scrambling to recover the stocks. Crouched behind the para-pet of a neighboring building,

> Superman's X-ray eyesight and super-acute hearing
> permit him to see and hear all that is occurring in
> the shabby room.

This intelligence prepares him for the attack on his Homer Ramsey identity in his office, not that the bullets wouldn't have bounced off him if he wasn't prepared.

Superman's task of going through the brokerage records might have been accelerated by X-ray eyesight, but he saves the power to use in eavesdropping joined with his equally super hearing. Super-man has already gone through nearly a year's worth of monthly adventures without using his enhanced sensory abilities. He contin-ues relying on his ability to leap (but not fly), repel murderous force, crush metal with his bare hands, outrace and catch cars, air-planes and bullets. The first account of his origin[208] portrays these alone.

Superman determines that an auto he sees speeding down below is not just a "joy-rider" but contains men fleeing with the abducted Senator Billingsley by glancing into the cab with his "X-ray vi-sion."[209] Superman's "marvelous super-sight" causes the roof of the car to open in a burst over the senator seated between two toughs

---

[207] Superman and the "Black Gold" Swindle, *Action Comics* 11, April, 1939 In Siegel and Shuster (2006: 146)

[208] *Superman* 1, July, 1939 in Siegel and Shuster (2006: 194-201)

[209] *Superman* 13, November/December, 1941 in Bridwell, ed. (1983: 82)

holding him dazed. Superman soon does rip open the top of the car and free the captive senator.

When Superman's X-ray vision tells him that the George Washington figure he finds himself beside within an impenetrable dome is authentic (he's carrying genuine Continental currency), General Custer, also present and authentic, asks him if he is "a circus performer?" "a trapeze artist?"[210] Superman then uses his X-ray vision to disrupt the generator circuits of the time machine aliens are using to extract famous people from history for a classroom demonstration, thus proving his boast that he has "strange powers...abilities beyond those of ordinary mortals."

Custer was generally correct in his assessment: Superman strongly resembles a circus performer in colorful costume standing out on the high wire. A pinup and a circular vignette that often appears in the corner of the early comic exhibits him bursting chains laced over his swelling chest, like a circus strongman. Once he stops traveling by leaps and takes to flying, his resemblance to a circus figure declines. He retains the identifying costume and in fact redefines it as the superhero costume with no suggestion of circus connections.

X-ray vision, which is soon accompanied and conflated with telescopic and microscopic vision, remains in the background of Superman's abilities. A repeated Kuda Bux-like performance, reading through bandages, would become routine for Superman's readers amid the very conspicuous feats of strength and endurance. During the 1950s the X-ray vision acquires a property that distinguishes it from all previous X-ray visions except that of Margot and Olga Mesmer: it becomes radiant vision with sighting capacities. Superman controls the destructive force of his vision and selects the aspect that will meet the occasion.

Superman can still look into a rampaging robot to determine who is controlling it, or detect a seeming spy within a trailer or tell whether a space ship is an illusion or not. Now he also can boil liquids, weld metals and set forests afire with his eyes. Unlike the

---

[210] *Action Comics* 399, April, 1971 in Bridwell, ed. (1983: 338)

visual X-ray vision this radiative accompaniment is as often destructive as it is beneficial.

An accident in space in 1957 increases the power of Superman's X-ray vision beyond his own volition: "Clear the street! Superman's X-ray vision is out of control!" exclaims the jagged text bubble of a fleeing pedestrian as the Man of Steel's eyes beam out burning rays.[211] The threat is eventfully brought under control. The following year Superman must desist from aiming shafts of energy at Brainiac.[212] The supervillain's self-enclosing force field reflects the rays back at Superman. In other 1950s threats a sinister puppet has obtained X-ray vision,[213] and elsewhere during the same period Superman's friend Lois Lane looks upon the world, and the superhero himself, with the painfully piercing eyes.[214]

The 1950s X-ray vision is conventionally illustrated with the same eye beams pouring from the face of Leo Brett and Jean Delorme at the beginning of the century. This is not the result of historic continuity: the rays have undergone a change in content while manifesting the same appearance.

The dangerous X-ray vision that seemed to be proliferating in Superman's world during the 1950s was an expression of the dangerous new radiation abroad during the age of atmospheric nuclear testing (1945-1962). The film version (1953) of H.G. Wells' novel *The War of the Worlds*, in which special effects wizard George Pal caused the Martians' heat ray to reduce a trio of pacifists to shadows on the ground, and mass flaming destruction in the cities and among the troops, brought that venerable fiction into the new era.

Superman's own encounters with the atomic bomb and nuclear radiation were precocious. A 1945 newspaper comic strip featuring Superman's facing the radiation from a cyclotron was suppressed at the request of the U.S. Department of Defense, as was a comic book story "The Battle of the Atoms"in which Luthor hurls an atom-

---

[211] The Man with Triple X-Ray Eyes, *Action Comics* 227, April, 1957
[212] The Super-Duel in Space, *Action Comics* 242, July, 1958 in Bridwell (1983: 266)
[213] The Super Puppet with X-Ray Eyes, *Superman* 109, November, 1956
[214] Lois Lane's X-Ray Vision, *Action Comics* 202, March, 1955

ic grenade at the superhero. The story was published in 1946, no longer a threat to national security.

Superman unfailingly resists the force of the atomic bomb whenever he encounters it, though it does knock him unconscious for thirty minutes. Superboy ingests a chemical cocktail that shows up as a nuclear blast on a fluoroscope screen and causes him to belch nuclear flares. Nuclear eyes never form a successor technology to X-ray eyes. After they are in the news, lasers are associated with the heat ray aspect of Superman's X-ray vision. Nuclear radiation and lasers do not include a way of seeing that also can be destructive, as X-ray vision does. The evolution of Superman's vision into a baleful force retains the X-ray identification.

As Superman became a figure of radio drama, movie serials, feature-length movies, television series, videogames, apps, and in correspondence with them children's toys, action figures, to name some of the media, his X-ray vision was displayed with all of the properties already associated with it. In the movie serials[215] and the first television series, *The Adventures of Superman,* the results of his X-ray vision were shown by moving from Superman's gazing eyes to the scene he saw after a shot of the intervening barrier. The 1993-97 television series *Lois and Clark* made the two reporters of the title, the second also Superman, the protagonists of the story. X-ray vision was looking through walls and objects in a burst to bring forward the action on the other side, except for the thickest wall, the one outside the ladies' locker room, which Superman blocked with his body when an X-ray fiend tried to aim his vision that way.

Superman's gallantry was sustained longer on television than it was in the movies. The 1978 *Superman,* his first feature-length film and the first in color, realized the aims of X-ray spectacles with an updating of the character when the Man of Steel told Lois Lane the color of her underwear. The underwear was not shown to the viewers as Superman saw it. The first X-ray eyesight of Action Comics had been adapted to the first putative use of X-ray spectacles. This first film encounter also allows critics to repeat that this cannot be X-ray vision since X-rays are not in color.

---

[215] *Superman* (1948) and *Superman vs. Atom Man* (1950) both 15-part serials

Other than routine spying out evil doings and unseen flaws, the Superman (and Supergirl) of the five movies a few times engaged in medical clairvoyance to locate a broken bone or assess internal injuries.

Tradition and innovation went hand in hand in the television series *Smallville*, which during its 10 seasons (2001-2011) was occupied with the teenage Clark Kent and the invention of his Superman secret identity, the high school years during the first four seasons and career beginnings thereafter. The future Superman acquired his powers and the ability to use them gradually, surprise by surprise. The episode entitled *X-Ray* (Season 1-Episode 4) enmeshed that ability in a plot that began in a bank robbery and culminated in the rescue of a girl entombed alive.

As in all individual X-ray vision Clark Kent's moments of seeing through barriers and into contents are signaled as private between the viewer and Clark by first showing his eyes then what he sees that no one else does. On a street of Smallville Clark sees into the clothing and skin of the fleeing Lex Luthor to detect in the greenish glowing outline of the skeleton several solid bodies, which turn out to be kryptonite embedded in the Luthor-appearing shape shifter.

This fluoroscopic X-ray subjectivity is repeated when Clark spots the same pattern of masses in the skeleton of a woman entering a store downtown. He also looks into a school locker and sees the money the shape-shifting student has hidden there, into a closed store where he sees the skeleton of a stuffed animal amid the greenish semi-transparent objects. In the culminating scene he stands up in a graveyard where he has fallen and sees inside a stone sarcophagus one of the skeletons moving slightly. He crushes the stone lid and raises Lana Lang who has been deposited there by the envious shape-shifter, who wants to enter her life.

This atmospheric greenish transparency of things is a computer graphics interface reconstruction of a set of objects or a person Clark has in view, to give the impression of transparency replacing solidity. It is not unlike the early cartoon representations of X-ray photographs or spectacle views of an entire scene, which implied what the solid scene looked like without presenting it.

Two other forms of X-ray vision come before Clark's eyes. As they are scaling ropes in the gym Clark suddenly sees the facial and upper torso musculature of the friend on the rope beside him. Shocked, he falls to the floor and looking up sees through the wall to the locker room where the dressing girls are in bras and briefs, and, in a moment that registers only on Clark's face, where Lana drops her towel, only shoulders laid bare for the viewers. In the Superman comic books where the X-ray spectacles were sold this event never was represented.

Standing in the high school hallway Clark gazes at the woman he believes to be the shape-shifter to no avail, yet he is able to will his vision to penetrate the metal door of the locker, emblematized by a circular opening at the center of the metal sheet. This appears to be a disavowal of the power-dominance gaze of the male teenager, but not of gym wall-penetrating voyeurism.

Clark puzzles out loud to his parents about the repeated involuntary and then voluntary acts of X-ray seeing. His step-father holds out a closed fist and asks him to say what is concealed inside. There is no X-ray transparency and Clark says he does not know. Jonathan Kent shows him he is holding a pocket knife. Stage performance and magic manual X-ray vision, the guessing games and blindfolded readings, also are disavowed, of course within the screen performance of the *Smallville* series.

Clark sees first in X-rays then he finds that he can look in X-rays but not that he can gaze in X-rays. The limits of his growing power are psychohistorical: they form at the division between past X-ray visions and present ones with an insistent recurrence to the look of actual X-rays.

A phenomenology of advanced sight gone wrong is sometimes encompassed in fictionalizations of new technologies. Ultrasound, for instance, spawns the occasional ambiguous use. In an episode of the television series *Alphas* (Syfy channel, 2011) a character, a gynecologist named Dr. Curran who lacks optic nerves, is able to form ultrasound images of people and their interiors, which are shown onscreen. He generates the high-frequency sound, which he can direct to cause structures to break up and collapse. This pairing, beneficial and destructive, is his only power. Dr. Curran is an ultrasound Xylope. The facility certainly helped him in his career examining women non-invasively.

These figures with extraordinary sight parallel technologies. They stand in a central place from which their view is taken, taking advantage of a shower of radiation or generating their own.

15

X-RAY SPECTACLES

As dreams and fears of imaginary X-ray spectacles arose during the early years of the 20th century, devices that would pass as X-ray spectacles were being patented. They eventually emerged as an X-ray vision purchasable by the public.

George W. MacDonald of Philadelphia, Pennsylvania wrote in the specifications letter for his Optical Device that it was an improvement upon optical instruments, useful as a toy, a scientific demonstration and "an attractive advertising medium".[216] A piece of transparent material etched or pressed with closely parallel lines was to be placed over a hole between two small pieces of opaque material. Anyone looking through the hole at, say, a human hand, would see the hand become transparent while objects seem to be extended in the field behind it.

MacDonald, the owner of an optical instruments firm, proposed to sequester the common light effect of diffraction as an illusion when focused on a single object. Any object seen in strong light through a transparent diffraction grating will seem to be extended out of itself due to the rays of object-reflected light crossing each other at different angles (interference) before reaching the eye. The object will seem to have a dark, solid interior surrounded by an outline of its shape flushed with the separated colors of the spectrum.

---

[216] George MacDonald, Optical Device, U.S. Patent Number 839, 016 Patented December 18, 1906

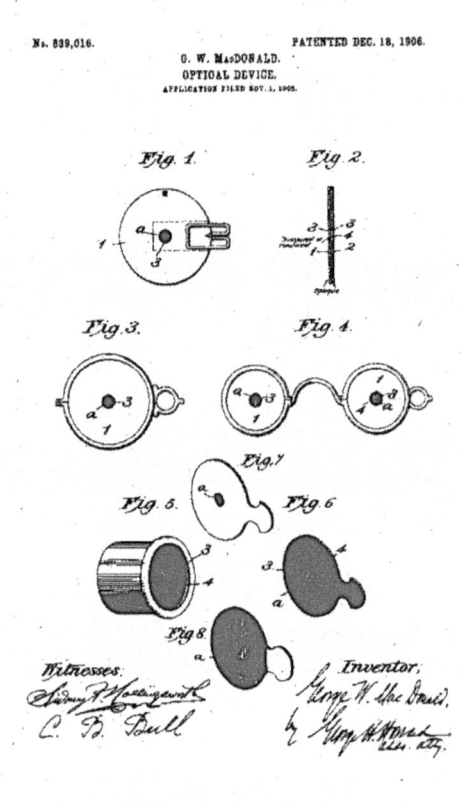

25. Optical Device. U.S. Patent drawing. Fn216

The spectroscope, which interposes a prism or a finely lined diffraction grating between a light source and the eye, had long been in use to break down a light source into colored bands that revealed the chemical composition of the light source. MacDonald's design strongly resembles cheap monocular spectroscopes constructed to give a general reading of light's composition. They can be a toy for anyone looking through the peephole and not concerned about the chemistry of the source. The viewer then would see spectral doubling of the object and be amused by the sight. That is the feature MacDonald patented. He doesn't use the word "spectrum" in his specifications.

He doesn't use the word "X-ray" either. One and a half years later Fred J. Wiedenbeck of Glenwood, Wisconsin does evoke X-rays in the letter of specifications for his patent of an "Optical-Illusions Device."[217] This device is a cigar-shaped wooden tube intended to serve as a monocular, one end open and the other fitted with a disk-shaped eyepiece of two layers of cardboard pierced in the center by a pinhole covered by a portion of bird feather fixed over the hole.

> Such being the construction of my improved optical illusion device it is held to the eye of the user, he preferably facing a strong light and looking toward a light background as the sky, and a solid object like a pencil or the fingers viewed through the device. The object viewed will then appear translucent, and in the case of the fingers the darker or solid portion will suggest the bones of the hand. And the general effect will be in fact, as far as what is seen is concerned, like an effect produced by the familiar and scientifically truthful X-ray device.

Wiedenbeck specifies a feather, shaft included, as the diffraction grating because he apparently is conscious of MacDonald's previous patent using a finely lined transparency. The device is "dependent for its operation upon the phenomenon of interference of rays of light passing through a grating of finely ruled lines." Using a finely lined bird's feather as the diffraction grating is Wiedenbeck's improvement. Shaping the tube like a cigar, and placing a seal resembling a cigar band over the joint, is an advertising component of the design.

The spectroscopic diffraction interference device has become an X-ray metaphor. The visual displacement of an object sighted

---

[217] Fred J. Wiedenbeck, Optical-Illusion Device, U.S. Patent Number 914,904 Patented May 25, 1908

through the spectroscope is a stand-in for the interior of the object inside its shell configurated by an X-ray photograph or a fluoroscope. An X-ray skirt, where the shadow of legs is projected onto a cloth screen, is more like an authentic X-ray image than the Optical-Illusion Device, which depends upon the distorted crisscrossing of the light rays reaching the retina.

The X-ray identification of this imagery was sustained for decades by not representing it other than through the device itself, a simple toy perhaps given away as a product favor. It is a play X-ray surrogate something like the metal detector packaged as a geiger counter at mid-century.

The spectroscopic interference toy using lined transparency or feather was patented several times in different forms during the rest of the century. Charles Raizen called it plainly an "X-Ray Simulating Toy" and geared it to offer both feather-based X-ray illusions and microscope illusions in an alernating frame.[218] In 1971, the "Sea Monkeys" entrepreneur Harold N. Braunhut entered an "Optical Device for Simulating Optical Images"[219] and followed it two years later with "An Optical Toy for Simulating Stereoscopic X-Ray Images."[220] Braunhut included the manufacturing process in another X-ray toy patent in 1990, and that same year added a camera for photographing simulated X-ray images, another of his enduring contributions to American fun.

Braunhut's 1971 patent was the first one to include both spectacles and a monocular, this time using a stressed thermoplastic material (diffraction grating replica) as the medium. A replica sheet was the active heart of a simulation that only resembled X-ray images because the patentee said so. The immemorial X-rayed hand, a verbal example in the earlier patents, appeared as an image in Braunhut's specifications.

---

[218] Charles Raizen, X-Ray Simulating Toy, Patent Number 2,537,332, Patented October 24, 1950

[219] Harold N. Braunhut, Optical Device for Simulating Optical Images Patent Number 3,592,533 Patented July 13, 1971

[220] Harold N. Braunhut, An Optical Toy for Simulating Stereoscopic X-Ray Images, Patent Number 3,711,183 Patented January 16, 1973

PATENTED JUL 1 3 1971   3,592.533

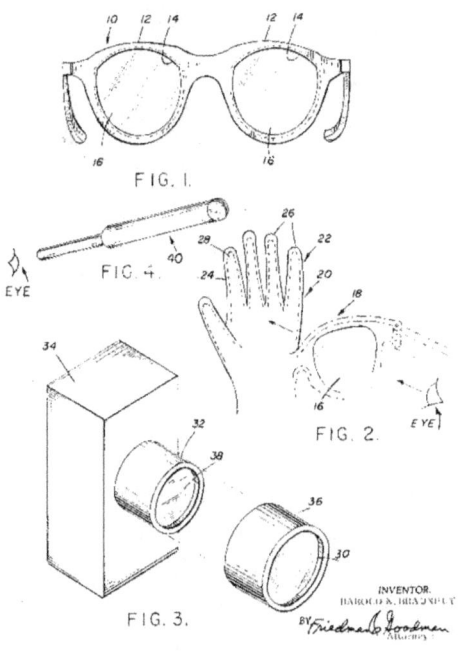

26. Optical Device for Simulating Optical Images
U.S. Patent drawing. Fn219

Braunhut's 1971 patent came at the end of the great age of X-ray spectacles, which apparently had proceeded without the benefit of patents. From the early 1960s onward X-ray spectacles were advertised for sale in comic books and popular science magazines, often depicting the "sees thro' clothing" trope that was uniformly absent from all of the patent specifications.

27. X-Ray Specs, *Popular Mechanics*, October, 1964

The advertised spectacles were a pair of plastic eyeglass frames filled with cardboard inserts with colored spirals painted on the outside, a pinhole poked through the center of each. Looking through the glasses at a person conveyed a doubled body streaked with iridescent greens and red. It didn't matter that the effect was achieved by chicken feathers crossing the pinhole. The X-ray reach of the specs was to be believed by the object of viewing, as suggested by an advertisement in the October, 1964 *Popular Mechanics*.

Other X-ray specs ads depicted the silhouette of a hand within its outline, a reversion to the first X-ray photograph. All of the ads disclaim the use of X-rays, if only by calling the specs an "optical illusion" and placing "see" in quotes. They all imply what the leering male face above beholds. Or can pretend he beholds to the person before him. The text reads "his" clothes but the figure is clearly that of a woman in a skirt.

The X-ray specs are a revival of the imaginary X-ray spectacles of the 1900s materialized in the modest technology of the patents and aimed at the burgeoning youth market of the 1960s. No cigars are implied. A number of reminiscences of that period recall the pretense that they might allow boys to look into girls' clothes, the danger of its being true, if only because of the girls' reaction to the voyeuristic intent present with or without X-ray spectacles.

During the early years of the X-ray specs period a film comprising their vision was released, *X: The Man with X-Ray Eyes* (1963). Starring Ray Milland, nearing the end of his career, and directed by Roger Corman, close to the beginning of his as a director of low-budget shockers, the film centered around the experiences of a doctor who acquires X-ray vision, and full radiesthesia by testing vision-enhancing eyedrops on himself.

Dr. Xavier first becomes aware of the voyeuristic potential of his new vision at a dance where we, the audience see only the bare shoulders of the naked dancers but Dr. Xavier must see more. He intervenes critically during surgery to save the life of a young girl whose internal organs he can clearly see, and is accused of malpractice by the chief surgeon. The arc of his career from this time onward is like that of an addict. He continues to administer the drops to extend his vision, and in resisting a friendly attempt at sedation he accidentally brings about the death of a supportive colleague. Now a fugitive, he falls into the clutches of a sideshow promoter, who hawks his skills in on-the-spot diagnosis.

Escaping that life, Xavier makes his way to Las Vegas where his eyes, now concealed by shades, give him an edge in the slots and at the blackjack table until the casino management notices. He wanders into a tent revival meeting in the desert, where the preacher shouts out the enjoinder from Matthew 5: "if thine eye offend thee, pluck it out." He removes his glasses, revealing eyes that have become jet black, and does as the preacher commands.

Roger Corman has at different times confirmed or denied the rumor repeated in reviews of the film that there was a further scene: his eye sockets empty, Xavier cries out, "I can still see!"

Xavier's point of view post-eyedrops is at first like that of diffraction grating spectacles without an object, a spectroscopic blur. Lying in bed, his eyes bandaged, Xavier sees his associates through the bandages in the manner of Kuda Bux. At the party and in a few other scenes Xavier sees bare legs and shoulders corresponding to the clad partygoers others see. It is implied that he can see others entirely without clothes, but as with all other X-ray eyes forays the whole naked body remains off-limits. During this first phase of his vision enhancement Xavier is able to focus into the flesh, bypassing the skin, and see stretches of tissue he then interprets medically. He performs the clairvoyance show of reading writing on a piece of paper placed against his blindfolded face and then answers the question posed by gazing into the pockets of the inquirer.

Confronted by the barker who's now funding his operation, Xavier sees an X-ray photograph of his skull slightly offset from the outline of flesh streaked with red and turquoise. His look into the leg of a woman injured in a carnival ride detects not an X-ray of the bones but a diagram of fractured bones. Other examinations of people's interiors are anatomical drawings.

All of these scenes are surrounded by a brownish ring like the focusing reticule of a microscope, usually with colored streaking around the rim. After his first eyedrop treatment he adopts sunglasses to which protective shades are added as his condition progresses.

Each time he removes the glasses his blue eyes appear streaked and bloodshot, then blank white, and finally in the last scene all black, an extreme of dilation. From the passenger side of the car a friendly colleague is driving, he looks out upon a spectroscopic cityscape and finally the neon signs of Las Vegas also fractioned into primary colored glowing lines. The system credited with these effects is Spectrama, which seems to be a cinematic version of X-ray specs. That remains Xavier's subjectivity for the remainder of the movie.

The film runs the gamut of X-ray eyes pseudo-technologies and postures. It is a catalogue of the appearances projected for them. Once Xavier (and the audience) is placed in this viewpoint he must experience the entire history of X-ray eyes until the spectroscopic vision becomes inescapable. His permanent and inescapable X-ray eyes are a look into the near future of movies, when similar destabilization of color values and figure integrity reflects drugged and paranormal states.

There long has been talk of a remake of the movie. Today's vastly superior special effects would be hard pressed to overcome the need for an other-than-X rating if the lucrative pre-teen crowd is to be attracted. This in addition to the real monotony of universal nakedness or skeletonization of others. A pornographic movie entitled *The Girl with the X-Ray Eyes* (no date) confirms this.

In the years since *The Man with the X-Ray Eyes* first appeared other men, women, boys and girls have been endowed with the ability to see as Dr. Xavier did, usually on television and as the result of obtaining "genuine" X-ray spectacles rather than by means of eyedrops. *The Kid with the X-Ray Eyes* (1999) does not attempt the Spectrama vision of Xavier's doom, but does include a skeleton in place of a fully clad person, men reduced to their underwear and the bare-shouldered women clearly not in evening dresses.

The Boy with the X-Ray Eyes (1999), a straight-to-video teen adventure film, also dispenses with any significant female involvement other than as objects of adolescent gazing, but does center around spy comedy as two teens who have innocently removed X-ray spectacles from a government laboratory try to evade the evil scientist who wants to sell the weaponized glasses to the Russians (now, of course, ruthless capitalists).

The second 1999 film introduces a new X-ray vision surrogate technology to take the place of the old diffraction grating-bird feather material. These films would not have been made if the possibility of seeing through clothing were not hinted at and then made family friendly. Infrared filters accomplish this by filtering out all radiation except infrared rays emitted by body heat under conditions of reduced light.

The dark body within an envelope X-ray surrogate developed by MacDonald and Wiedenbeck (and the makers of the X-ray skirt) is revived by means of a technology which does outline the body within the clothes. The early 21st century saw the SONY Corporation alter the Nightvision feature of a video camera to prevent it from being used to see through clothing during the day. Filters were offered for sale and advice was given on Internet sites on modifying existing cameras for daytime infrared inspection activities.[221]

Rival secret agents vie to recover "top-secret X-ray vision glasses" that have accidentally come into the possession of a "ditzy strip-

---

[221] Thompson (2004)

per" in *The Girl with the Sex-Ray Eyes* (2007) shown on the poster art wearing a pair of mail-order X-ray Gogs. The Sex-Ray Eyes are in fact infrared-enabled.

The absorption of the X-ray specs into the nostalgia industry was made complete with the trademarking of an X-ray vision design based on the 1960s comic books ads, for printing on a range of clothing and accessories from thongs to sunglasses. X-ray spectacles (X-ray Specs, X-ray Gogs) continue to be available at a much higher price and in forms that suggest they are now part of the nostalgia market.

An image of the spiral-centered eyeglasses with the word mark "Scam-Co's X-Ray Vision" adjoining a text like that of a comic book ad offering the glasses for $1.00 has been trademarked (2005 and still live) with the U.S. Patent and Trademark Office for use on t-shirts, thongs, underwear, eyeglasses and many other accessories and articles of clothing.

The X-ray glasses ended out adorning what they were expected to allow the wearer to see and see-through.

X-ray spectacles are a cultural system that admits of a range of distortions of vision on the verge of beholding a never to be attained personal nakedness of others. Nakedness is sought; nudity is achieved. No X-rays are present, except in name.

16

APPLICATIONS

Reason Warehime, a soldier in the U.S. Army, was stationed with other enlisted men in a trench 2500 feet from the April 25, 1953 Shot Simon atomic bomb test site in Nevada. After the initial pressure of the heat wave compressed his body, the light reaching him was so intense that he could see the bones of the hand he held up "like they were being X-rayed."[222]

28. Shot Simon, 17 kt.., Nevada Test Site, 18 August, 1957

---

[222] Interview quoted by Halton (1998: 49)

Other soldiers, similarly stationed and equally unprotected, had the same experience at nuclear and thermonuclear tests, in the American desert, Australian outback and at sea near Pacific islands.[223]

These men came as close as possible to having X-ray vision in the natural world, though it was intense light and not X-rays that showed them the bones inside the skin. The X-rays and other radiation generated in the blast passed through skin, bone and eyes without a screen to register the shadows they cast.

The ill-fated soldiers were the opposite of shamans and witches worldwide, who had the X-ray vision of being able to look into human bodies without implying that X-rays were involved. "Fictive sensory paths of the agentive visual type" in the phrase of the philosopher Paul Bloom.[224]

The shamans are the opposite of the soldiers, initiators rather than receivers, having visions of invented light rather than seeing a genuine illumination, a skill rather than a one-time experience. The soldiers were raising their hands to shield their eyes, not because they planned to see the bones shadowed by the rays. The shamans on the other hand were trained visionaries looking into people for a purpose and for a reward. Both saw as with X-rays, whether that designation was imposed or not. Both were agents of X-rays.

The magical practitioners saw into the flesh of others under circumstances just as specific as surrounded the soldiers. The X-ray environment unites them in a common act of visual agency. A light peculiar to the occasion, emitted and accepted, makes the insides register.

During the !kia healing ritual of the !Kung people, of the Kalahari, the celebrants walk on fire, see with X-ray eyes and over great distances.[225] They are like Kuda Bux and Superman, and like the atomic veterans at the same time. Heat and light are simultaneously

---

[223] James Yeatts at Operation Tumbler-Snapper in Nevada (de Groot 2005: 242); Ken McGinley at Operation Grapple near Christmas Island (Roff 1997: 52)

[224] Bloom (1999: 233)

[225] Katz (1976: 287)

present in the body. The X-ray occasion is created by placing the body where it can burn and see.

Marjorie Mandelstam Balzer found that one of the Siberian Sakha healers she consulted detected an "abnormality" she knew was within her.[226] "X-ray vision?" she placed in parentheses after the statement. An interest in X-ray vision in Russia is especially strong among those with "shamanic spiritual traditions."[227]

The interpretation of shamanic insights is conditioned by the independent Russian occult background bridging personal consultations with shamans. A longtime student of Siberian shamanism and its contemporary adaptations links shamanism with her own Western body through X-ray vision. Soviet and post-Soviet proposals to remake humanity now can include the long-persecuted Siberian shamans. Russia's own girl with the X-ray eyes, Natasha Demkina, is a present-day shaman subjected to psychic investigation.

The Hopi had an "eye society," *poovost* or *poswiwimkyana*, "men of X-ray vision" an association to which all shamans belonged.[228] It was defunct by the time an ethnographer interviewed its surviving members in the 1890s. Just as X-rays were being discovered the opportunity to learn what "eye" the society members were using faded away. Some reflection of that vision, sufficient to associate it with X-rays, was preserved in tales of a shaman using a rock crystal to uncover evil men.

The eye society men cast a glance over the pueblo and saw evildoers. Potential evildoers were aware that they were under the watchful eyes of the *poovost*. Except for his unfortunate reliance on internal radium for his ability to see through walls, Lucien Delorme might have been a member of this society. The dwellings of the Hopi pueblos were connected with each other by walls surrounding a central plaza. X-ray vision was the knowledge held by all inhabitants that someone committed to maintaining public order was attentive to the sights and sounds. Awareness of someone with this

---

[226] Balzer (2006: 98n15)

[227] Kravchuk (2010: 268nc)

[228] Malotki, Gary and Knorowski (2006: xxviii-ix)

power next door would have placed a constraint on the activities of the residents of a Paris lodging as well.

X-ray art, interiors of animals and humans, of the inhabitants of dwellings and vessels, is on view in the drawings, engravings, murals and carvings of many peoples.[229] The significance of the exposure, as far as known, varies from one representation and one people to another. Mayan and Olmec murals include a shaman's profile visible inside the outlined features of an animal mask.

29. Drawing after Mural 1 figure, Oxtotitlan Cave, Guerrero, Mexico

This drawing after Mural 1 over the opening of Oxtotitlan Cave in Guerrero, Mexico places a human profile wearing a nose plug and ear ornament inside the mask of an owl as he sits upon the altar his arms extended in a dance gesture of flight. The shaman's eye is clearly visible, as searching and vigilant as the eye of the owl he impersonates. A masked figure from the Mayan temple at Yaxchilan is eye level with staring eye of the jaguar mask he is shown within.[230]

---

[229] Lommel (1967: 129-33)
[230] Spinden (1913: 22)

30. Drawing after masked figure, Yaxchilan, Chiapas, Mexico. Fn230

The blindfolded performer of X-ray eyes is also letting the audience know he is watching them become aware of his eye upon them, only it is deflected to a commonplace gesture, words upon a board or a page from a newspaper. His eye, like the shaman's, is alive within the enclosure, ready to act upon what he sees. The audience is made aware of this.

Members of the Gelede, a society of masked dancers of the Yoruba, themselves do not claim X-ray vision, but they know that witches are capable of developing that among other powers.[231] The "witches" are people with a malefic force within themselves who can emulate the witchcraft of the white man without the technological trappings: X-ray vision without an X-ray machine, long distance talk without a telephone. These "mothers" (originators) may not intend to exercise the powers they develop spontaneously, out of an unadmitted resentment or jealousy. It is up to the Gelede to channel the powers of the mothers during nocturnal dances criticizing morally errant village residents. The ability of mothers to look into other people is then turned to the benefit of the whole social group. The poovost of the Hopi and the Gelede of the Yoruba have this purpose. The Gelede, still functioning after whites brought the X-rays, sought a way to harness that witchcraft before it spread out of control.

---

[231] Henry Drewal cited by Thompson (1974: 98)

X-ray vision also is among the strange powers acquired by an Akan witch of Ghana.[232]

> She sees herself lying in her bed while she floats around her body, and she may take short trips outside her room and see that she can penetrate walls and fly outside her room. She flies around her neighborhood and discovers that not only can she see in darkness, but she can also see the sleeping bodies of her neighbors, and the walls of their homes pose no barrier to her.

In this case there is no white technology serving as the source, and no Gelede society to try to reroute it beneficially. There is only Gabriel Bannerman-Richter, who makes a distinction between witchcraft, in which the actions of the agent are voluntary, and spirit possession, where they are the work of spirits entering a person.

This newly acquired X-ray vision is neighbor-specific; only the walls of the nearby houses, not clothing or skin, are meaningful to see through. It is not the urbanized, anonymous X-ray vision of a superhero that can evolve into a harmful force. The quality of the ability, in tandem with flight, bends her (witches are female only) upon mischief already perking up in her since she has purchased the witchcraft for a purpose.

The secrets of houses are fair game for the Akan witch; the secrets of bodies are distressing. Bannerman-Richter recounts a story that circulated during the 1940s of a pair of Ghanaian sisters who went to the Nzima region of Togoland to buy trade enhancement witchcraft. The sorcerer who sold them the charms warned them that there would be consequences. As they boarded the bus to return each sister suddenly discovered that she could see the other's insides. Alarmed and not at all curious about the ability, they hurriedly gave the charms back to the provider.

---

[232] Bannerman-Richter (1982: 43)

X-ray vision defines the limit of what information can be sought about others. Even those seeking a magical edge in trade stop at spying inside the fabric shield. The boundary lines that are teased by X-ray spectacles and observed by Superman remain intact for witches. X-ray vision that investigates the social world veers away from the body insides.

Looking inside the body can make a shaman into a vampire. Wana shamans of Sulawesi gain X-ray vision only during the performance of ritual, allowing them to visualize the contents of a vessel without the vessel, such as water standing in space.[233] If they possess this ability all the time they lust to taste body organs, especially the placenta of a gestating infant. Medical diagnosis does not result from seeing inside. Hunger does, and hunger must be satisfied. If the vampire can fly, another possible result of the magic that leads to X-ray vision, then he or she has a greater chance of fulfilling the new hunger. They become a cause of affliction which more restrained magical practitioners must resolve.

The Akan purchasers of charms shy away from X-ray vision of the body when it becomes possible. The Wana shaman become vampire is lured into gorging himself on the riches of the body, becoming an explanation for miscarriages and stillbirths. X-ray vision grants a freedom not usually possible or even conceivable without that sudden flash of insight. This is the temptation hovering beside the potion of eyedrops and the charm of X-ray spectacles, held out tantalizingly before the eyes of anyone who imagines ahead of sight. The prospect of sex without resistance, or free food and unimpeded access to the private quarters of others all hang just on the other side of the barrier that X-ray vision magically penetrates. The white man's X-ray is a skill that can be transferred to the adept, not an apparatus requiring a power source, film and a processing laboratory.

The Ojibway of Ontario, Canada had their own traditional technique that allowed those bent on malice to cause illness by damaging the interior of the body. A person with a grievance, he need not be a witch of shaman, drew a picture of the intended victim in "X-

---

[233] Atkinson (1989: 97)

ray style" then "violated" it.[234] Just representing the exposed insides of a body made it vulnerable to mischief. It is possible that the many X-ray figures of animals and humans from the cave art of the Paleolithic to the present day had a similar purpose. The shaman's eye inside the mask looks out upon a world he can control. He himself is controlled by being in the figure.

The twentieth century Canadian painter of Ojibway origin, Norval Morrisseau, made extensive use of X-ray representations of humans and animals in his brilliantly colored canvases, drawings and prints. He fell afoul of Ojibway elders who complained that he was disclosing communal secrets in his works. X-ray vision was, after all, reserved for special occasions. The men who saw their hands X-rayed in nuclear light could not disclose that sight until decades later.

---

[234] Vecsey (1983: 109)

17

KNOWLEDGE OF THE BUTCHERED MAN

The anthropologist Thomas Gregor asked Mehinaku Indians of Suyapuhi (a village near the Xingu River, Brazil) to "draw the inside of a man."[235] Gregor himself titled these "X-ray drawings." There was no tradition of X-ray drawing among the Mehinaku. Gregor wanted to see graphic expressions of Mehinaku ideas of human nature and sexuality within the frame of the body. He knew that the Mehinaku believed that certain plants eaten as food become semen and other food plants become feces. Their drawings would trace the route the food follows through the body because they had specific traditional knowledge of the body's interior.

The Mehinaku often named their settlements after important events. Suyapuhi was named for the killing of a Suya Indian 40 years earlier. Out of curiosity, the Mehinaku communally decided to examine this man's interior. There were strict prohibitions against opening the bodies of Mehinaku dead, and an enemy's body had not previously fallen into their hands. Under the guidance of a Wauru Indian (of yet another group, not enemy) a few of the men undertook the task.

All of these men had died by the time Gregor stayed at the village. Their knowledge had entered into oral tradition and integrated with what was learned from butchering monkeys for food. Passageways between the sac where the food resided and the area of the genitals and the anus gave evidence for their beliefs about the course of food. The Mehinaku made X-ray drawings for Gregor illustrating this process. They had acquired X-ray vision.

---

[235] Gregor (1987: 85)

This was X-ray vision impelled by the energy paths they observed within the butchered man. They had no knowledge of X-rays, and Gregor did not try to impose it on them. Like many other anthropologists and art historians he used X-ray drawing as a label, and in this case the label structured his own act of data collection. He saw the depicted body as a result of the Mehinaku's own history of looking and seeing cast within a framework he provided.

Other peoples had arrived at representations that could be termed X-ray drawing from the viewpoint of observers from other cultures. Many of them were recording their knowledge of animal and human body insides gained from butchering and warfare. The Mehinaku did not make any such record but retained the knowledge in oral tradition and were prepared to inscribe it when Gregor asked them to.

Their vision most closely resembled the knowledge of physicians who took part in dissections and brought that knowledge to the diagnosis of complaints. They looked into a body for the anatomical root of the pain or disease symptoms visible on the outside. This knowledge was increasingly codified in drawings and models as the technology of print grew more sophisticated and pervasive. It still involved carrying the image of the disposition of skeletal bones, vessels and organs in the mind's eye.

With all the use of the phrase "X-ray vision" as a means to look into bodies, it has been uncommon in writings, technical and popular, by the medical profession and about their work. X-ray vision or X-ray eyes did not describe what the doctors saw in the body any more than it described what the Mehinaku knew and saw. Physicians learning anatomy could be encouraged to think of this knowledge as a form of X-ray vision where the phrase is familiar from vernacular usage.[236] There are many different pedagogical approaches to gaining and retaining anatomical knowledge, but X-ray visualization is not first among them.

Does the medical view of the body, or the scientific view of space and matter, all susceptible to X-ray investigation, constitute a default cognition which is replaced by the common idea of X-ray

---

[236] Whiten (2006: 1)

vision? Is the way trained doctors see the body, astronomers see the depths of space, microbiologists see the world under the microscope, metallurgists see the interior of fatigued metal, are they cognitively all forms of X-ray vision *ex litteris*? Knowledge of the body once butchered?

This is not a question in the light of the various manifestations of X-ray vision reviewed so far in this book. There does seem to be cognitive object, what I have called an image-idea that filled the space so opportunely named X-ray vision after X-rays were discovered. The phantasmagoria, the earth vision of the dowsers and medical clairvoyance occupied this space but none of them filled it exclusively.

The X-ray vision space is for everyone who expressed seeing into objects: a culturally conditioned space not dependent upon what it "actually" looks like inside an object, what genuine X-rays show there, if they can show anything. Exceeding X-ray limitations, seeing in colors, seeing diagrammatically, seeing across distances and into minute quarters, seeing thoughts and emotions, that fills the chronicles of X-ray vision.

Studying the application of the X-ray vision phrase to classify the looking of cultures other than those that used X-ray technology alerts me to the presence of this image-idea. X-rays circumscribe the vision they fill in. The Mehinaku look into an interior that we would not expect to see using X-rays. Superman and The Man with the X-ray Eyes do not see the interior X-rays shadow out, but a culturally meaningful set of insides. X-ray vision is the drive and thrust of the looking, not the vision. The cultural space which X-ray vision names is broader and more diverse than X-ray vision. It is not the same as an X-ray of the body.

The early Indian anatomists, Atreya (600-700 BCE) and Sushruta (400 BCE) counted many more bones in the adult human skeleton (360-300) than are recognized by Western anatomists today (206). This discrepancy is reconciled with contemporary skeletal anatomy because the Indian anatomists counted teeth, nails and cartilages, separated the bony processes from the bones, and added a few bones for the sake of preserving symmetry, for instance a third phalanx in the thumbs.[237] They did practice dissection, by allowing a body to decay in running water, and made direct observations of the body's interior during surgery. The image of the body they had before their eyes was the result of investigation and practical manipulation within their own experience which can be cast in terms of other medical systems only by ignoring the forces they believed to be operating in health and illness.

Contemporary X-ray vision would find two phalanges in the thumb where the X-ray vision of Atreya and Sushruta would find three. Both visions are the result of anatomical texts and demonstrations/dissections backloaded into the investigative eye. The texts, which were the main transmission of anatomical information, provided the culturally conditioned template for what should be seen in the open body. X-ray vision as it developed historically in Europe and America was a departure from both directly observed anatomy and the anatomy interpreted from X-ray photographs. It was an assertion that looking directly into the body (or any other enclosed vessel) would not require the training of a doctor or other specialist in that interior.

X-ray vision as practiced in the twentieth century thus violates the traditional ways of acquiring and acting upon that knowledge of body internals. It was not an ability that doctors imagined they had, even fictionally, because it was a throwback to a way of looking that preceded X-rays and was not even supported by knowledge gained by dissection or opening the enclosure. X-ray vision dared to look back and see what was already known appropriate to the worldview of the looker. Western classical anatomy was book-ridden and subject to personal authority for millennia, the hints of knowledge gained from dissection leaking down from one practitioner to another. Revisions in the book knowledge of anatomy that formed the view of the practitioners were slow in coming and contested, yet ultimately they formed part of the textbook view. The discovery of X-rays created a double for the older vision drawn from tradition and not experience. X-ray vision was fixed upon the skeletons of the phantasmagoria and the earth of the dowsers, the person inside the clothing but not the naked body.

---

[237] Hoernle (1984: 115)

Most at stake in looking deep inside or far outside is the risk of incomprehensibility. A fixed format that realizes a worldview overcomes that risk in advance. No one who acquired X-ray vision, prior to Olga Mesmer and Superman, could live with that ability for long. Only when part of someone with a suite of other powers, forming a well-organized body, could X-ray vision be sustained. The expedient of removable X-ray spectacles allowed for a temporary X-ray vision. Permanent X-ray vision was parochial and only found what might be expected. When Dr. Xavier began to see the cosmos he tore his eyes out. The superhero does not open him or herself to the vast ocular perspective. There is law and order to maintain here on Earth.

I do not know of a fiction in which the possessor of X-ray vision sees an interior he cannot comprehend. Like the people staring into the heads of spouses and intimates they know only too well what they see. When Maria Trompeyeva sees the void inside the aristocracy she rationalizes and embraces it.

Let me postulate, however, that a woman gains X-ray vision and sees this inside a friend.

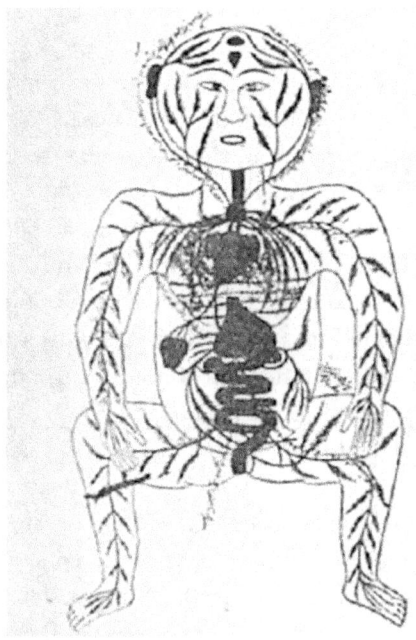

31. Anatomical System. Cowdry (1921: Plate 1)

If she expected to see a textbook anatomy diagram, her bewilderment is palpable to those of us seeing her vision from the outside. Her X-ray vision is not indigenous to her own culture. She has not been provided with an advance on what she will see by her own background. If her culture has a provision for X-ray vision its activation may look very different from this. Or she may have the kind of X-ray vision that admits of acupuncture.

X-ray vision always has a use, and certainly carrying forward the acupuncture meridians in the form of a present insight can be one of those uses. But what if this X-ray vision comes to her with no cultural preparation. Within the context of traditional Western X-ray vision it would have to mean that the acupuncture meridians are the inside of the body. It then constitutes a confirmation of someone else's symbol system as a reality of the body, like the anatomical diagrams Dr. Xavier and Clark Kent see inside other people. Only those diagrams are the only ones the English physician and the Kryptonian illegal immigrant are supposed to see.

Our X-ray vision endowed woman comes up against the interior concreteness of an alternative body, one that she might accept as such but does not think is real as the anatomy of Western medicine is real inside the body exposed to her. The same might be said of the interior of the earth or the depths of the ocean.

Making visions that transcend sight accessible to those who are not experiencing them is always problematic. They end out being flattened into representation or telegraphed in a code that secures their remoteness. It is good to have a tool like X-ray vision to facilitate that transition, or to imply that the received object is in truth miraculous and beyond our senses. X-ray vision is a temptation to reverse engineer toward but never to a much, much more powerful experience than the one before us. The knowledge of the butchered man is the knowledge of his being butchered, not taking part in the act. Probably not.

The old template realized in X-ray vision is realized in opposition to the overwhelming sensory loads of plunging into a body or coursing into the depths of the earth. X-ray vision synthesizes a clean cut into a resisting body that will release fluids, guts, dust, and make sounds obstructing clean sight. The twentieth century development of high powered weapons that send projectiles tearing flesh into anonymous bits, and the progressive use of explosive devices in conflicts with a similar result is in direct opposition to the X-ray vision model with its staging of a cleanly anatomical viewpoint.

X-ray vision certainly does endure under these conditions, and provides an alternative history to views of bodies and enclosures being entered non-destructively. A new form of X-ray vision not given that name has come into practice in recent decades. This form, virtual anatomy, maintains the same relationship with texts, diagrams and authority that X-ray vision does, but proceeds with an energy like that of a projectile, even taking the same course as a projectile into the body or other enclosure.

Virtual anatomy is on exhibit in the television crime drama series *Crime Scene Investigators (CSI)*. From its debut in 2000, CSI has followed a team of Las Vegas police forensic investigators as they apply scientific methods and instruments to identifying murder victims and their killers. Now in its 12th season, CSI airs in dozens of countries and has served as a model for spin-offs, derivatives, novelizations, games, and scenes in other television series and films. In this hour-long (in fact, 42 minute-long) police procedural the investigators have more diversified roles than they do in real-world police homicide investigations. They gather and analyze evidence; they also secure crime scenes, seek out and apprehend the accused, and serve as expert witnesses.

Except for one character who was a qualified physician, the CSI's on the show do not perform postmortems, which are in the hands of a pathologist serving as medical examiner. The cause of death examinations usually take place in the examination room at police headquarters where bodies of those who appear to have suffered foul play are brought. These are the circumstances for the CSI version of X-ray vision. The camera trained upon the wound or visible affected body area plunges into the skin and follows a course through the viscera and whatever body system is affected to the location of the mortal event. This course can mimic the path of a bullet or other projectile along whatever route it did follow, or that of electric current coursing along a nerve, or the molecules of a caustic substance through the circulatory system. A legally acceptable cause of death is turned into a mini-drama of the anatomical body.

The human anatomy the projectile travels through in these CSI movements is not a photographic body. Actors and/or made-to-order manikins are used for the outside view, for picturing the general state of the body of the deceased and the location of trauma. The virtual anatomy shots are computer generated graphics of precisely the areas entered and traversed by the deadly object or force. All the blood vessels, nerves, muscles and glands in an area are shown in colors distinct enough to contrast and more in keeping with the coloring of contemporary anatomical illustration than actual anatomy.

The makers of CSI virtual anatomy work in the stage tradition leading from the Parisian *Theatre du Grand Guignol* of the early twentieth century, where at least 5 different formulations of stage blood were used to achieve effects of dripping, spreading and spray that could reach the audience. Instead of bracing themselves to meet the wave of gore, the audience of CSI is drawn in to the mass of gore instructively labeled, more like jury members than a theatre audience.

That motion is secured by animation: the perimeters of the open body move past the viewer as quickly as an amusement park horror ride, not quite as fast as a speeding bullet, more at the speed of a knife blade. The sound accompanying the motion suggests physical contact between a moving object and a fluid, but like the motion itself slowed down to be identifiable. The slap of a broad front entering a viscous surface leads to slurping sounds of travel through thick fluid rather than a high-pitched bullet shot covering all. The sound is synthesized, or recorded and then mixed to project an object event that projects the actual body acoustics of murder, familiar to no one. The sound is as true to the bullet entry and travel as the shapes and color are. This makes a complete package.

The CSI version of X-ray vision takes place at most once during an episode. Instead of privileging the viewer, like "true" X-ray vision, the CSI form needs no lead story of special eyedrops or extraterrestrial origin to make someone capable of seeing into bodies.

The noisy dive into the flesh is an assertion of expertise that makes the cause of death plain to everyone. It is a collective visual inscription of how the death happened according to the evidence collected and analyzed by the CSI's and framed by the verbal discourse of the medical examiner. Rather than the vision entering the body, the body clusters around the thrusting pivot of the vision.

The use of this expedient has remained during the recent seasons of the series, and is present in the spinoffs (*CSI: New York* and *CSI: Miami*). *CSI: New York* favors a luminous networks of vessels and organs enclosed by a cloudlike transparent blue body shape, a floating virtual body whole corresponding to the corpse studied by the medical examiner. There is no dynamics of movement into this body. It is a collective X-ray vision, a transparent world form of the virtual vision, again without a superhero author.

The *CSI: New York* virtual anatomy is manifest within the drama as a product of an unexplained technology. It is a step away from the CSI thrust in being a virtual reality heuristically viewed by the characters rather than simply configuring their cumulative reconstruction of the death event. The image on screen is a model created by a private corporation, *Zygote Media Systems*, that licenses similar anatomical models for educational and entertainment use.

These television virtual anatomies realize X-ray vision as a simulacrum of three-dimensional vision. Both the driving projectile and the floating body seem to occupy a space with extension inside the context of the screen drama. They fulfill the need of the scripted action to dramatize a search within that space for causes and information contained by the body. They are both examples of the increasing tendency to see X-rays as displacement in space equivalent to perspectival forms of the object entered. X-rays photography of the patient's body had already undergone this extension in Computer Assisted Tomographic Scans (CAT Scans) which creates stacks of "slices" of the subject's interior. The virtual bodies of television fiction are the old X-ray vision catching up with and advancing on CAT Scans without the medical need to attain strict correspondence with an individual.

A collaboration between anatomists, biochemists, medical illustrators at Rochester Institute of Technology in 2006 fashioned "an interactive voyage into human anatomy."[238] The viewer wore a pair of goggles that gave the sense of spying the pancreas within the layers of the abdomen. The skull and the molecular level of the DNA molecule also dispose themselves in view. This body is not photographic, and does not correspond directly to a genuine patient. The illustrated body has an interior space generated by the act of looking.

The fictional virtual bodies for television audiences and the three dimensional one for single viewers created by scientists are equally distant from X-ray vision as a look directly into or onto a single body for diagnostic or any other purpose. The old fictional and surrogate modes of X-ray vision did not achieve a look into an actual body either. These newer fictions are still the same kind of fiction, dependent upon a template that satisfies X-ray vision and virtual reality image-ideas from dancing skeletons and glowing water sources onwards.

The Rochester interactive voyage does include an element present in other X-ray vision attempts: looking inside the body is also looking inside the cells down to the microscopic level. External barriers are also barriers of scale and distance. The ability of X-rays to visualize the unseen world inside things is identified with the ability of the microscope and the telescope to visualize the very small and the very remote. Whatever is not immediately present to the eyes is included in the template of what X-ray vision can accomplish.

Using the Advanced Photon Source, the most powerful source of coherent X-rays available, a group of scientists were able to engineer a computer program that interprets the X-ray backscatter from a material down to the level of separate magnetic domains and shows a visual image of the field.[239] An article on the imaging system compared it to Superman's X-ray vision.[240]

---

[238] Virtual Human Body, *Science Daily* [online], January 1, 2007.

[239] Tripathi, et al. (2011)

[240] Like Superman's X-ray Vision, New Microscope Reveals Nanoscale Details, *ScienceDaily,* August 9, 2011.

Superman had (or has) both X-ray vision and microscopic vision: an imagistically logical pairing. What use is it to see into a body and not make out the details of bodily processes, down to the cellular level if necessary? The dichroic microscope, using X-rays and not visible light, processes information about magnetic flux carried by the scattered X-rays. That will be useful analyzing the information storage capabilities of materials to improve data processing memory.

The scale maintained and collapsed by light imagery is equally preserved in X-ray imagery. Superman had telescopic vision as well. The Chandra X-ray Telescope, an earth-orbiting satellite, records X-ray emissions from sectors of space and plots the location of the source, its shape, relative speed and distance. Besides accomplishing their intended purpose, these two technologies together amount to genuine X-ray vision. One uses an artificially generated source of X-rays and the other is highly sensitive to existing natural ones.

The dangerous, subjective form of X-ray vision fictionally and metaphorically bestowed on individuals is not realized in the dichroic microscope and the X-ray observatory. Another type of X-ray vision is realized here. The implementation of these technologies certainly will not end the centuries long quest for penetrating sight brought into focus by the discovery of X-rays. It does open the imagination beyond the somewhat narrow terms of that quest and its fictive realizations. Diffraction grating and infrared sensors will still be used to simulate personal X-ray vision, and project the sexual politics of that assumed fact.

Now we can see with X-ray eyes, like the eyes of the Xylope because the microscope and telescope use sensors opaque to light, into which X-rays can pass. Not "we" in the sense of all of us. The pictures will be delivered in correspondence to the magnetic domains and intergalactic X-rays sources as X-ray photographs of bodies and shoes and paintings have been, for the viewing of those they might concern, as art if that seems to be the case.

In the Standard Model of contemporary physics the elementary particle the photon is the carrier of all forms of electromagnetic energy, including visible light, infrared light, X-rays and gamma rays. At rest a photon is without mass, travels at the speed of light, and as visible light has just enough energy to excite the molecule of a photoreceptor cell in the eye. The frequency of the wave it carries accounts for the energy of a photon. All the forms of X-ray vision, from visible light and infrared surrogates to X-rays themselves are expressions of photons carrying waves of different frequencies. The interior of the hand is both seen and X-rayed in a nuclear blast because all forms of photon are rushing toward the viewer.

The physics of the photon, both particle and wave, forms a unified field theory for X-ray vision. The eye receives all of these energy states but only renders visible those states of a frequency that can excite molecules in the photoreceptors. The lower frequencies, including light itself, excite the molecules into a temporary energy state from which they return, releasing a lesser amount of energy in the form of a nerve impulse. The higher frequencies, including X-rays, are likely to decompose the receptor molecule. It is separated into its constituent elements, and the tissue it forms decays. The blind people into whose eyes Edison beamed X-rays detected the inner light of photons released by excited tissue. The optic nerve transmitted a signal like light to the brain. It only was the energy of decomposing tissue released into the optic pathways, the signal occasioned by the radiation.

Do the blind experience migraine auras as the sighted do? Migraine auras are colorful weaves of light that stand before the eyes with or without an accompanying headache. The lights are generated within the optic nerve. Do the same signals generate flashes in the eyes of the blind?

Sight itself is a field of energy converted into electrical impulses by the rods and cones in the retina and resolved into signals by the brain. Forms of radiant energy other than light stimulate activity in the optic nerve that forms sight and if repeated a type of vision. The dichroic microscope and the Chandra observatory operate analogously to the eye and brain that ultimately receive and interpret their images. This is X-ray vision using real X-rays.

18
LIGHTNING GIVES COP X-RAY VISION!

X-ray vision struggles to be born in us. If we cannot spontaneously see with X-rays we say or think we do, or that someone does. Or that something does. This narrative has many branches, recalling past X-ray visions and imagining those that already exist if only it is recognized. No longer is X-ray vision in the future; it is now, in the present of all previous beliefs come together.

Officer John Elvar was struck by lightning while using an experimental digital surveillance camera and the software was transferred to his brain, giving him X-ray vision.[241] Learning to sleep with transparent eyelids and letting his beard grow because he couldn't find his chin to shave, Elvar also had to overcome his wife's reaction to his wince every time he saw her. It is not explained exactly how his eyesight enabled him to solve the murder of a judge and close down a brothel. The women working in his office have been lining their underwear with Mylar to block his prying, to no avail, he hints.

Elvar's tabloid reported sensory enhancement is induced by the interaction of the oldest power-source in mythology, lightning, acting upon a new human-absorbable technology, digital camera software. Force plus surveillance equals X-ray vision: the old formula repeated in renewed terms. The same X-ray vision as always, at first a hazard and an embarrassment then a heroic tool, and finally cagily voyeuristic.

X-ray eyes also are a means of expressing male discomfort with female discomfort with the male gaze, in the work settings where that gaze is steadily deployed. In several cases, again reported by

---

[241] Vincent (2005)

tabloids, women in a Liverpool housing office and in a Paris welfare office object to a male co-worker they say has X-ray eyes enabling him to see through their clothing.[242] Hundreds of these "idiot feminists" go on strike, unsuccessfully demanding his transfer. The article is from the viewpoint of the male protagonist in his bewilderment at the women's beliefs. Why is he being singled out from the other men in the office? The articles implicitly raise that question without answering it.

Elvar embodies X-ray vision subjectivity worked into in a social world. He and his office worker counterparts face joint action by women in their daily vicinity. The women's collective reaction, a stereotyped subjectivity, comes into play for the man with "genuine" X-ray vision and for the men ascribed X-ray vision by the women. In none of the articles do the women directly state the reasons for their actions. The rumor of X-ray spectacles allegedly caused women to resort to shielded undergarments immediately after X-rays were announced. The office replaces the street as the zone of encounter.

Western media, or at least a few elements within it, became aware that the village of Grokhov, Russia had been blanketed by radioactive fallout as a result of the Chernobyl meltdown on April 26, 1986, and every person in the town subsequently developed X-ray vision.[243] They stared out of all-white, opaque eyeballs (shown in a photograph) through walls and metal casings. Investigative journalist William Raynes learned of the events in Grokhov only in 1994, when he obtained Soviet documents relating efforts to learn how the radiation caused the change in the villagers' vision. The Soviets plan to militarize their findings must have been undone by the 1990 collapse of the Soviet Union and the independence of the Ukraine, where Chernobyl, and presumably Grokhov, is located.

Raynes' book, *Chernobyl Today*, is not in evidence today, nor is the X-ray vision of residents of any contiguous village part of the dossier of the short or long-term effects of the Chernobyl disaster. Grochow, Poland, far from Chernobyl, was the site of a skirmish

---

[242] Berger (1988) and (1992)

[243] Fine (1994) and (1995). The same article repeated.

between Russian and Polish troops in 1831, and Grochow was the name of a farm where Jewish fighters against the Nazi advance gathered in 1940, but the name has no resonance in X-ray annals. Radioactivity, like lightning, was known to induce superhuman powers, which the Soviets would have been eager to employ.

A faint memory of the Xylope, Lucien Delorme and Dr. Xavier lingers in the eyes of the people of Grochov, but unlike these predecessors they do not disclose much of what they see. The condition of some people living in the vicinity of Chernobyl, sickness and death from thyroid cancer, was the main discernible result of the most serious civilian nuclear accident. Soviets, the Cold War, nuclear fears and military mysticism combined to give the old X-ray vision a new platform in Grokhov.

The X-ray glasses were being improved upon as well. "Sex-driven" 76-year old inventor Bill Walton intended to market his battery-powered X-ray glasses in America after he was released from a French jail.[244] He was doing time for ogling and pursuing women in a Paris supermarket while wearing his invention. A photograph of the man's face extends dotted rays from his black framed glasses toward the behind of a woman in a tight black negligee. The comic book glasses of an earlier era were just a simulation; Walton's long years of work and $20,000 had yielded a working prototype the principle of which he would not discuss.

Excessive success also bedeviled the makers of See-All-X-Ray-Specs, who were besieged by 33,000 Japanese men demanding their money back.[245] Instead of the skin beneath the dress the Specs gave the eager viewer a sight of the bones, nerves and blood vessels.

This tale of thwarted peeping, the most consistent theme in the history of X-ray vision, is ever available to provide a note in an existing political scandal or synthesize a new one. President Bill Clinton, in the throes of his sex-scandal driven impeachment, was given a pair of X-ray glasses by a thoughtless aid, with embarrassing consequences. President George W. Bush took out a pair of comic book X-ray spectacles to inspect the Secretary of State, Condoleeza

---

[244] Layton (1990) and Ross (1992). Almost the same, a few details vary.

[245] New X-ray Specs are Too Good! *World Weekly News*, January 15, 2002

Rice, precipitating an angry meeting behind closed doors and contradictory statements by government officials.[246] The specs work too well and deny the desired view, or are discovered and denounced. They are the vision men aim at women, and that the powerful aim at their subordinates, formalized in so many other ways, in this case not allowed to work freely.

The X-ray eyes theme is combined with many others to build tabloid stories. It does not change. It is attached to current news in the company of other themes, without surrendering the sexual surveillance anxiety at its root. That theme only is relinquished when X-ray eyes are a metaphor for a way of looking divorced from present circumstances and events, a proposed new way of looking that might fulfill the promise of X-rays become a property of human eyes.

A writer on chess strategy, Patrick Wolff, offers his readers Superman's X-ray eyes aimed at the chess board.[247] By envisioning the moves and captures possible once a piece has been moved a player can see through the current position of a piece to the outcomes of subsequent moves. It is a mapping strategy expressed by other chess writers over the years without resorting to Superman or his extraordinary vision. The novice player's confidence is boosted by thinking of the board array in purely visual terms before recurring to the movement differences between the pieces. This is X-ray vision not looking back upon its sexual politics.

X-ray vision can be recruited metaphorically to stand in for a new way of looking at how we see. An entire chapter of Mark Changizi's popular book on the cognitive science of vision is devoted to X-ray vision.[248] Changizi contends that humans already have an X-ray vision superior to Superman's because it is not obstructed by lead. We cannot see directly through solid objects as the Man of Steel can. We resolve figures glimpsed on the other side of a field of separate pieces like blades or stalks narrower than the distance between our eyes. By staring at a drawing of a human face outline

---

[246] Kennedy (2004)

[247] Wolff (2001: 119)

[248] Changizi (2009: 49-108)

covered by grass stalks we can see only the stalks or only the face and thus we can see through a barrier to the object on the other side.

It takes some visual habituation to make Changizi's X-ray vision work. He achieved this discipline and came to recognize it while playing video games that reward sighting an enemy lurking in a thicket while keeping the thicket in view. It is a neuroscientific X-ray vision replacing the cinematic circumstances of its birth. At first it seems just to be a loose application of the metaphor to an otherwise explicable optical field.

Changizi has, however, extended X-ray vision in tandem with technological developments that have remade the eye of the screen viewer into an X-ray eye. I account my difficulty working his exercises due to my habit of looking at screens passively to view films and videos or follow the cursor as I type and maneuver files. I do not play video games and lack their visual discipline of looking into as well as at the screen. Three-dimensional screen motion arriving in video games on axes across and into the screen is a similar development permitted by digital processing of enormous quantities of visual information. X-ray eyes are embodying one more generation of technology.

Changizi sees this X-ray eyes view as the accompaniment of natural selection in favor the ability of early hominids to see prey in glades and while stealthily approaching through the brush. Binocular and color vision were good for spying the fruit that could provide the vitamin C hominids might not get if animal prey was too large and too elusive. Before an evolutionary scenario is constructed from this, it is good to recur to the technology that now governs the view that is not vital to anyone but the very few hunter-gatherers who remain on the planet. We can't know if they had the X-ray vision films and videos have now given to us.

Against this abstract or at least disinterested X-ray vision is a trend toward making a view achieved through a change of light into a competent surrogate. Bird feathers and diffraction grating X-ray oculars were the first attempt, successful because they defined X-ray vision without X-rays almost from the start. The recruitment of infrared radiation as the initiator of X-ray vision is step toward simulation that requires more investment from the would-be viewer.

The purchaser of comic book X-ray specs was taking part in group illusion that updated the early twentieth-century X-ray specs joke as a fixture of youth culture retained in nostalgia. Looking through the cardboard lens holes gave the viewer a shifted, opalescent vision and alerted others to the peering intent, underlining it with their reaction, like that of Condoleeza Rice to President Bush's use of the specs in the tabloid story. Pretending that the glasses really could look into clothes only as far as the skin was training for some in negotiating sexual interest and reputation, as the tabloid stories also reflect, and formulations like Changizi's reflect that by not reflecting it.

Cardboard X-ray specs were not replaced by anything because the specific type of looking they represented was otherwise technologized in youth culture. Boy/girl with the X-ray eyes movies and television episodes replaced man/woman with the X-ray eyes. If you wanted to let someone of the opposite (or same) sex know you were looking at them with X-ray eyes you would not say so in those terms. Greater sexual freedom and a growing number of portable video recording devices made the X-ray eyes interchange seem monolithic and archaic. The perpetual throwback to the first years of X-rays, reputed spectacles and defensive undergarments was finally over. The cop in the latest tabloid acquisition of X-ray vision received it from a surveillance camera.

The marginal uses of consumer goods become widely known when a manufacturer decides to remove them from the market out of embarrassment and anticipated liability. The Sony Corporation recalled the NightShot version of its Handycam video camera in 1999 after an article in the men's magazine *Takarajima* described how to make it able to allow the user to see through clothing during the day.[249]

There is no public record of how many cameras were returned. A talk show host joked that Sony recalled the cameras-and raised the price. Sony did modify the feature on later NightShot

---

[249] Patton (1998)

Handycams to confine them to night use, and in their advertising emphasized the domestic and unaggressive uses of the equipment. An urban legend had been born. As the twentieth century progressed oral tradition on how to overcome the camera's disability crystallized in booklets[250] and instructional videos easily found on the Internet. The sexual surveillance substrate of X-ray vision had been reincarnated.

Instructions on how to modify the Handycam generalized the process to other cameras. The basis of the small, portable video cameras' potential was a CCD, or in later models a CMOS, that received the light of the image received by the lens and translated it into electronic impulses to be recorded for later reconstruction and replay. These receptors were highly sensitive to the infrared end of the spectrum, to such a degree that the cameras included an infrared reflector between lens and chip. Infrared signals caused blur and color shift in the recorded image if they reached the receptor. The Sony NightShot Lux 0 allowed videography in darkness because the reflector could be manually raised away from the light path with a switch on the camera body. The infrared rays proceeding from anybody emitting heat were recorded. Many cameras also had an infrared emitter to enrich the scene night-watched.

To keep the infrared capabilities engaged for daytime viewing only required switching to night mode and placing on the lens a filter than blocked visible light in favor of infrared. In full daylight the long infrared rays in sunlight are abundantly reflected by light bodies and fully absorbed by dark ones. Using filters and infrared sensitive film still photographers have long taken photographs giving scenery an appearance of unnatural contrast. Those images ceased to be a novelty long ago. The infrared sensitive video camera could detect ink writing beneath an ink blot by recording the differentially greater absorption of rays by the letters underneath. And it could presumably see through clothing by recording the rays reflected by the skin beneath the fabric. [1]

That was the subject of the Takarajima article and the selling point of all subsequent infrared X-ray camera arrangements. Sony's

---

[250] Thompson (2004)

withdrawal of the cameras was a marketing strategy that boosted their mystique and challenged video hobbyists to make the modifications. Warnings by the authors of manuals that the X-ray vision of the cameras should only be used "appropriately" also signaled that it had the potential not to be used in that way. Statements that it was difficult to obtain an image of skin even through thin fabric, and that did not represent the contours of the skin underneath. The few copies of X-ray vision infrared photographs printed in books and posted on the Internet also would be either a warning or challenge to those pursuing the view promised by all X-ray vision.

The photographs and videos most revealing of the vision being sought and accepted were not shots of nudes, which could easily be dismissed as fakes. They were hazy frontal photographs of manikins first dressed and then X-rayed to disclose a pair of sharply delineated nipples and the lettering on a card that had been inserted down the front of the dress. They defined X-ray vision for all its history. The message could be read but it blocked the skin that was the objective of the visionary.

Many X-ray vision infrared cameras are offered for sale through independent dealers found by Internet searches. Instructions for refitting existing equipment, for modifying mobile phones or any other device that can be used for clandestine picture taking. They offer the X-ray way of looking mechanized, digitized a la mode, and only to be fully attained with the purchase of the instrument. X-ray vision is then whatever is visible through that ocular. No complaint would be "appropriate" if the desired vision doesn't present itself to the eye. The only standard against which to measure the result is what already has been seen without X-ray vision, or imagined according to a fiction, and that result is subject to all the influences that launched us on the quest for X-ray vision in the first place.

A beryllium window supplied by NASA that sends out photons corresponding to the X-rays it receives? The dichromatic glasses supplied with the *Superman Returns X-Ray Vision Activity Book*? They also allow someone to see as X-rays would, or should allow, but do not.

REFERENCES

Atkinson, Jane Monnig. 1989. *The Art and Politics of Wana Sham an-ship*. Berkeley: University of California Press.

Babbitt, Edwin Dwight. 1874. *The Health Guide: Aiming at a Higher Science of Life and the Life Forces*. New York: E.D. Babbitt, D.M.

Balzer, Marjorie Mandelstam. 2006. Sustainable Faith? Reconfigur-ing Shamanic Healing in Siberia, 78-100 IN *Spiritual Transfor-mation and Healing*, ed. by John D. Rosi-Chioino and Philip Hofner. Oxford: Alta Mira Press.

Bangs, John Kendrick. 1902. *Olympian Nights*. New York: Harper and Brothers.

Bannerman-Richter, Gabriel. 1982. *The Practice of Witchcraft in Gha-na*. Elk Grove, California: Gabari Publishing Company.

Barker, George F., trans. and ed. 1898. *Roentgen Rays*. New York: American Book Company.

Barrett, William F. 1897. On the So-Called Divining Rod, *Proceed-ings of the Society for Psychical Research* 15: 130-383.

Barrett, William F. 1913. The Psychical versus the Physical Theory of Dowsing, *Journal of the Society for Psychical Research* 16: 43-48.

Barrett, William F. and Theodore Besterman. 1926. *The Divining Rod: An Experimental and Psychological Investigation*. London: Me-thuen and Co.

Berger, Joe. 1988. Put on Some Shades, Buster! *Weekly World News* December 20: 17.

Berger, Joe. 1992. Put on Some Shades, Buster! *Weekly World News* January 9: 33.

Bhana, Deeksha, et al. 2009. *Student's Guide to the Law of Contract*. Kluwer.

Biddle, Lisa. 1902. Invisible Spectacles: A Phrenological Experi-ence, *The Phrenological Journal of Science and Health* 114, 1: 226-27.

Blanco, Ramiro. 1900. El mundo de los rayos X, *Revista Contempo-ránea* 118: 364-71.

Bleiler, Everett F. 1998. *Science Fiction: The Gernsback Years*. Kent, Ohio: Kent State University Press.

Bloom, Paul. 1999. *Language and Space*. Cambridge: M.I.T. Press.

Boatright, Mody. 1963. *Folklore of the Oil Industry*. SMU Press.

Bossalino, D. 1906. Sur la visibilité des rayons X, *Archives Italiennes de Biologie* 46: 68-72.

Bourrelier, Paul Henri. 2008. Portrait d'un Dreyfusard: Gaston Moch, combattant de la paix, *Bulletin de la SABIX* [online] 42: 75-91.

Brandes, G. 1896. Ueber der Sichbarkeit den Roentgen Strahlen, *Sitzungberichte der königliche Preussiche Akademie den Wissenschaften zu Berlin* 24: 547-50.

Brett, Frank W. 1899. Professor Frank W. Brett and His Son: An Editori al Sketch, followed by Conversation with Frank W. Brett, M.D., *The Coming Age: A Magazine of Constructive Thought* II, 5: 449-55.

Bridwell, E. Nelson, ed. 1983. *Superman From the 30s to the 80s*. New York: Crown Publishers.

Bulgakowa, Oksana. 2005. Eisenstein, the Glass House and The Spherical Book: From the Comedy of the Eye to a Drama of Enlightenment. *Rouge* [online]

Burns, P.W. 1998. *Television: An International History of the Formative Years*. London: The Institution of Electrical Engineers.

Changizi, Mark. 2009. *The Vision Revolution: How the Latest Research Overturns Everything We Know about Human Vision*. Dallas: BenBella Books.

Cowdry, E.V. 1921. Ancient Chinese Anatomical Charts, *The Anatomical Record* 22:1-26.

Crookes, William. 1897. Sur la relativité des connaissances humaines, *La Revue Scientifique (Revue Rouge)* 7: 609-13.

Crosthwaite, C.H.T. 1869. *Notes on the North-Western Provinces of India, by a District Officer*. London: W.H. Allen and Co.

Crosthwaite, C.H.T. 1896. Röntgen's Curse, *Longman's Magazine* 28: 469-84.

Crosthwaite, C.H.T. 1897. Thakur Pertab Singh: A Tale of an Indian Famine, *Blackwood's Edinburgh Magazine* 162: 28-54.

Dacynski, Vincent. 2004. Firewalking Kuda Bux-The Man With X-Ray Eyes. www.amazingabilities.com/amaze/6.html

de Groot, Gerard. 2005. *The Bomb: A Life*. Cambridge: Harvard University Press.

Deindorfer, Robert. 1949. Pieter's X-Ray Eyes, *Life* April 25: 2+.

Duff, Edward Macomb and Thomas Gilchrist Allen. 1902. *Psychic Research and Gospel Miracles*. New York: Thomas Whittaker.

Edwards, Frank. 1966. *Strange People*. New York: Lyle Stuart.

Eisenstein, S. M. 2009. *Glass House: Du Projet de Film au Film comme Projet*. Paris: Les Presses du Reel.

Emory, William H. 1848. *Notes of a Military Reconnaissance from Fort Leavenworth, in Missouri, to San Diego, in California*. Washington: Wendell and van Benthuysen.

d'Espérance, Elizabeth. 1897. *Shadow Land, or Light from the Other Side*. London: George Rodway.

Evans, Constance Mary [Mairi O'Nair]. 1934. *The Girl with the X-Ray Eyes*.

Fine, Karl. 1994. The Town with X-Ray Eyes, *World Weekly News*, January 4-11: 66 and September 19, 1995: 32.

Feijoo, Benito Jeronimo. 1739. *Demonstracion Critico-Apologetica del Theatro Critico Universal*. 4 v. Madrid: Los Herederos de Francisco del Hierro.

Fischer, Ottakar, June Barrows Mussey and Fulton Oursler. 1931. *Illustrated Magic*. New York: The Macmillan Company.

Fleitz, David. 2002. *Louis Sockalexis: The First Cleveland Indian*. Charlotte: McFarland.

Foveau de Courmelles, François. 1898. Visibilité des rayons X par certains jeunes aveugles, *Journal d'Hygiène* 23: 173-75.

Glasser, Otto. 1993. *William Conrad Röntgen and the Early History of the Röntgen Rays*. Novato: Norman Publishing.

Godber, Noel. 1931. *Amazing Spectacles!* London: John Long Limited.

Goodman, David C. 2002. *Power and Penury: Government and Technology in Philip II's Spain.* Cambridge: Cambridge University Press.

Goupil, M.A. 1898. Lucidité: Experiments du Dr. Ferroul, *Annales des Sciences Psychiques* 6: 139-50; 193-99.

Gregor, Thomas. 1987. *Anxious Pleasures: The Sexual Lives of an Amazonian People.* Chicago: University of Chicago Press.

Grasset, Joseph. 1897. L'experience de lecture à travers les corps opaques, *Annales des Sciences Psychiques* 7: 321-26.

Grasset, Joseph. 1908. *L'occultisme Hier et Aujourd'hui: Le Merveilleux Prescientifique.* Coulet et Fils.

Griffith, George. 1896. A Photograph of the Invisible, *Pearson's Magazine* April: 235-40.

Grimshaw, Robert. 1911. The "Telegraphic Eye," *Scientific American* 104: 335-36.

Guillaume, Charles-Edouard. 1896. *Les rayons X et la photographie á travers les corps opaques.* Paris: Gauthier-Vilau et Fils.

Guthrie, Christopher. 2010. Socialism in Microcosm: The Municipal Administration of Dr. Ernest Ferroul in Narbonne, 1896-1921, *Euro- pean History Quarterly* 40: 79-96.

Haines, Douglas Melvin. 2001. *Imperial Medicine: Patrick Manson and the Conquest of Tropical Disease.* Philadelphia: University of Pennsylvania Press.

Halton, Eugene. 1998. *The Great Brain Suck, and Other American Epiphanies.* Chicago: University of Chicago Press.

Hamilton, Edward. 1922. The Man with X-Ray Eyes. summarized in Bleiler (1998: 166).

Hartwell, David. 1997. *The Science Fiction Century.* London: Macmillan.

Hempel, Charles Julius. 1859. *A New and Comprehensive System of Materia Medica and Therapeutics.* Boston: William Radde.

Henriksen, Thormod and H. David Maillie. 2003. *Radiation and Health.* London: Taylor and Francis.

Hoernle, A.F. Rudolf. 1984. *Studies in the Medicine of Ancient India.* New Delhi: Concept Publishing Company.

Holbein the Younger, Hans. 1971 (1538). *The Dance of Death.* New York: Dover.

Héricoult, J. 1896. Le mois scientifique, *La Revue Hebdomaire* 1: 553-58.

Houdini, Harry. 1924. *Margery Exposed, Also a Complete Exposure of Argamasilla.* New York: Adams Press.

Houdini, Harry. 1953. *Houdini on Magic.* New York: Dover.

Huff, Richard. 2006. *Reality Television*. Greenwich: Greenwood Publishing Group.

Jacobs, Steven. 2007. *The Wrong House: The Architecture of Alfred Hitchcock*. Rotterdam: 010 Publishers.

Katz, R. 1976. Education for Transcendence: !kia Healing with the Kalahari !Kung, IN *Kalahari Hunter-Gatherers*, edited by R. Lee and I. DeVore. Cambridge: Harvard University Press.

Kennedy, Fred. 2004. New White House Scandal! *Weekly World News* December 27 8-9.

Kravchuk, Liudmila X. 2010. Activity of the Chinese Religious Movement Falun Gong in Russia, 258-70 IN *Religion and Politics in Russia: A Reader*. Armonk, NY: M.E. Sharpe.

Lavine, Matthew. 2008. *A Cultural History of Radiation and Radioactivity in the United States, 1895-1945*. Ph.D. Dissertation, Department of History, University of Wisconsin-Madison.

Layton, Dave. 1990. X-Ray Glasses Land Inventor in French Jail, *Weekly World News* April 3: 2.

Legendre, Charles Gilbert. 1735. *Traité de l'opinion*. Paris: Chez Briasson.

Lion, M.G. 1906. Estomac en sablier avec sténose medio-gastrique radiographie, *Bulletins et Mémoires de la Société Médicale des Hospitaux de Paris* 22: 110-19.

Lockyer, Norman, ed. 1896. Science in the Magazines, *Nature* 54, 1402: 454.

Lommel, Andreas. 1967. *Shamanism: The Beginnings of Art*. New York: McGraw-Hill.

McPartland, Donald Scott. 2006. *Almost Edison: How William Sawyer and Others Lost the Race to Electrification*. Ph.D. dissertation, Department of History, City University of New York.

Malcolm, Ian. 1907. *Indian Pictures and Problems*. London: E. Grant Richards.

Malotki, Ekkehart, Ken Gary and Karen Knorowski. 2006. *Hopi Stories of Witchcraft, Shamanism and Magic*. Lincoln: University of Nebraska Press.

Mannoni, Laurent. 2000. *The Great Art of Light and Shadow: Archaeology of the Cinema,* translated and edited by Richard Crangle. Exeter: University of Exeter Press.

Márquez, Enrique. 1991. El caso Houdini-Argamasilla: una información non veraz en la literatura parapsicológica española, *El ojo esceptico* 1,3 [online].

Mason, R. Osgood. 1901. *Hypnotism and Suggestion in Therapeutics, Education and Reform.* New York: Henry Holt and Company.

Miles, Elton. 1976. *Tales of the Big Bend.* "the author."

Millard, Bailey. 1916. What Is There in the Occult? *Illustrated World* 24, 5: 631-36.

Minkov, Svetoslav. 1965. *The Lady with the X-Ray Eyes*, translated by Kassimira Noneva. Sofia: Foreign Languages Press.

Moch, M. Gaston. 1897. Sur la relativité des connaissances humaines, *La Revue Scientifique (Revue Rouge)* 8,4: 104-8.

Moch, M. Gaston. 1898. Sur la relativité des connaissances humaines, *Bulletin de la Société Astronomique de la France* 12: 176-84.

Moch, M. Gaston. 1922. *Initiation aux Théories d'Einstein.* Paris: Bibliothèque Larousse.

Moch, M. Gaston. 1923. *La Relativité des Phénomènes.* Paris: E. Flammarion.

Moreux, Th. 1902-03. Radiations connues et régions inexplorées, *Revue du Monde Invisible* 5: 425-30.

Mosher, Leroy. 1905. *The Stranded Bugle, and Other Poems and Prose.* Los Angeles: The Times-Mirror Company.

Münsterberg, Hugo. 1913. The Case of Beulah Miller: An Investigation of the New Psychical Mystery. *The Metropolitan* 38: 16-18; 61-62.

Münsterberg, Hugo. 1914. *Psychology and Social Sanity.* Garden City, N.Y.: Doubldeday, Page & Co.

van der Naillen, Alexander. 1912. *The Strenuous Life Spiritual and The Submissive Life.* New York: R.F. Fenno & Company. [reference is to *The Submissive Life* separately paginated]

Nayak, Narendra. 2010. Exposing the "Miracle" of Blindfolded Sight: The Story of Ranjana Agrawal, *Nirmukta*, June 23 [online]

Owen, Alex. 2004. *The Darkened Room: Women, Power and Spiritualism in Late Victorian England.* Chicago: University of Chicago Press.

Panchadasi, Swami. 1916. *A Course of Advanced Lessons in Clairvoyance and Occult Powers.* Chicago: Advanced Thought Publishing Company.

Pang, Laikwan. 2007. *The Distorting Mirror: Visual Modernity in China.* Honolulu: University of Hawai'i Press.

Patton, Phil. 1998. Public Eye; Seeing in the Dark, *New York Times* December 3

Pear, T.H. 1935. Report upon and Examination of Mr. Kuda Bux's Claim to See with his Nostril, *Senate House Papers, University of London* [typescript].

Peck, W.H. 1897. Recent Contributions to Skiagraphy, *Medicine* 3: 1-2.

Peterson, West. 1949. This Man Has X-Ray Eyes, *Mechanix Illustrated* August 1949: 61-64; 152.

Podmore, Frank. 1908. *The Naturalisation of the Supernatural.* G.P. Putnam's Sons.

Price, Harry. 1939. *Fifty Years of Psychical Research.* London: Longman, Greens & Co.,Ltd..

Quackenbos, John D. 1908. *Hypnotic Therapeutics in Theory and Practice, with Numerous Illustrations of Treatment by Suggestion.* New York: Harper and Brothers.

Quackenbos, John D. 1916. *Body and Spirit: An Inquiry into the Subconscious.* New York: Harper and Brothers.

Ramakrishnan, Thuttakadu. 1896. *Tales of Ind, and Other Poems.* London: T. Fisher Unwin.

Rayon. 1897. Church, Science, and Natural Healing Methods, *The Metaphysical Magazine* 5,2: 124-34.

Rayon. 1900. *The Mystic Self: Uncommon Sense and Common Sense.* Chicago.

Remise, Jac, Pascale Remise and Regis van de Walle. 1979. *Magie Lumineuse: Du Théâtre d'Ombres à la Lanterne Magique.* Tours: Balland.

Rinn, Joseph Francis. 1950. *Sixty Years of Psychical Research: Houdini and I Among the Spiritualists.* New York: The Truth Seeker Company.

Robertson, Etienne-Gaspard. 1831. *Mémoires recréatifs,scientifiques et anecdotiques...* Paris: chez l'auteur.

Robertson, G.H. 1896. X-Rays and Blindness: Letter from G.H. Robertson, *The Electrician* 38: 291.

Roff, Sue Rabbitt. The Ghost of Christmas Past, *Bulletin of the Atomic Scientists*, September: 52-56.

Rosny, J.H. 1898. *Un Autre Monde*. Paris: Plon.

Ross, Bradley. 1992. X-Ray Glasses Land Inventor in French Jail, *Weekly World News* February 11: 21.

Sample, Ian. 2004. Visionary or Fortune-teller? Why Scientists Find Diagnoses of 'X-ray girl' Hard to Swallow, *The Guardian*, September 24 [online]

Shiers, George and May. 1997. *Early Television: A Bibliographic Guide to 1940*. Taylor & Francis.

Siegel, Jerry and Joe Shuster. 2006. *The Superman Chronicles, Volume 1*. New York: DC Comics.

Silverblatt, Irene. 2000. The Inca's Witches: Gender and the Cultural World of Civilization in Seventeenth-Century Peru, 109-30 IN *Possible Pasts: Becoming Colonial in Early America*, edited by Robert Blair St. George. Utica: Cornell University Press.

Singer, Natasha. 2012. Mission Control, Built for Cities. *New York Times* March 3: BU1.

Skolnick, Andrew. 2005. Natasha Demkina: The Girl with Normal Eyes, *The Skeptical Inquirer* 29, 3.

Smith, Bill. 2011. Guy Oliver Fenley: The Boy with the X-Ray Eyes, *Terrell County Memorial Museum News*. March. [online]

Smith, J.G. 1898. Some Cases Recorded in the "Annales des Sciences Psychiques," *Proceedings of the Society for Psychical Research* 7:115-18.

Spinden, Herbert. 1913. A Study of Maya Art: Its Subject Matter and Historical Development,*Memoirs of the Peabody Museum of American Archaeology and Ethnology*, Vol. VI.

Steadman, John F. 1962. Value of a Yankee Uniform, *Baseball Digest* March: 43-44.

Strosnider, J. Steve. 2002. *Tales from the Track: Stories from the Early Years of Racing*. Victoria, B.C.: Trafford.

de Téramond, Guy. 1914. *L'homme qui voit á travers les murailles*. Paris: Tallandier.

de Téramond, Guy. 1915. *The Mystery of Lucien Delorme*, translated by Mary Safford. New York: Appleton.

Tesla, Nikola. 1896. On Roentgen Rays, 267-71 IN *The Nikola Tesla Treasury* (2007). Radford, Virginia: Wilder Publications.

Thompson, M. Patrick. 2004. *The Mystery of X-Ray Cameras Revealed.* ADD Creations.

Thompson, Robert Farris. 1975. *African Art in Motion.* Berkeley: University of California Press.

Tripathi, Ashish, et al. 2011. Dichroic coherent diffractive imaging, *Proceedings of the National Academy of Sciences*, August 8, DOI: 10.1073/pnas.1104304108.

Vecsey, Christopher. 1983. *The Traditional Ojibway Religion and its Historical Changes.* Philadelphia: American Philosophical Society.

Vernier, J.P. 1975. The Science Fiction of J.H. Rosny the elder, *Science Fiction Studies* #6, vol. 2, pt. 2 [online]

Vincent, Doug. 2005. Lightning Gives Cop X-Ray Vision, *Weekly World News* May 23: 33.

Walsh, Anthony. n.d. Molly Fancher...the Brooklyn Enigma: The Psychological Marvel of the Nineteenth Century [inside.salve.edu/walsh/mollie.html]

Warnell, Phillip. 2008. *The Girl with X-ray Eyes.* The Mead Gallery.

Wehr, Hans. 1976. *Arabic-English Dictionary,* ed. by J.M. Cowan. Ithaca: Spoken Language Services, Inc.

Wells, H.G. *The Stolen Bacillus and Other Incidents.* London: Macmillan and Company.

Weston, Stephen. 1810. *Remains of Arabic in the Spanish and Portuguese Languages.* London.

Whiten, Susie. 2006. *Flesh and Bones of Anatomy.* Elsevier Mosby.

Williams, Keith. 2007. *H.G. Wells, Modernity and the Movies.* Liverpool: Liverpool University Press.

Wilson, Lyman P. 1922. The X-Ray in Court, *The Cornell Law Quarterly* 7: 202-34; 334; 51.

Wolff, Patrick. 2001. *The Complete Idiot's Guide to Chess.* Penguin.

INDEX